MOLECULAR BIOLOGY
INTELLIGENCE
UNIT

Programmed Cell Death in Protozoa

José Manuel Pérez Martín, Ph.D.
Centro de Biología Molecular "Severo Ochoa" (CSIC-UAM)
Universidad Autónoma de Madrid
Madrid, Spain

LANDES BIOSCIENCE
AUSTIN, TEXAS
U.S.A.

SPRINGER SCIENCE+BUSINESS MEDIA
NEW YORK, NEW YORK
U.S.A.

PROGRAMMED CELL DEATH IN PROTOZOA

Molecular Biology Intelligence Unit

Landes Bioscience
Springer Science+Business Media, LLC

ISBN: 978-0-387-76716-1   Printed on acid-free paper.

Copyright ©2008 Landes Bioscience and Springer Science+Business Media, LLC

Springer Science+Business Media, LLC, 233 Spring Street, New York, New York 10013, U.S.A.
http://www.springer.com

Please address all inquiries to the publishers:
Landes Bioscience, 1002 West Avenue, 2nd Floor, Austin, Texas 78701, U.S.A.
Phone: 512/ 637 6050; FAX: 512/ 637 6079
http://www.landesbioscience.com

Printed in the United States of America.

9 8 7 6 5 4 3 2 1

Library of Congress Cataloging-in-Publication Data

Pérez Martín, José Manuel.
  Programmed cell death in protozoa / [edited by] José Manuel Pérez Martín.
    p. ; cm. -- (Molecular biology intelligence unit)
  Includes bibliographical references and index.
  ISBN 978-0-387-76716-1 (alk. paper)
  1. Protozoa--Physiology. I. Title. II. Series: Molecular biology intelligence unit (Unnumbered)
  [DNLM: 1. Protozoa--physiology. 2. Apoptosis--physiology. QX 50 P438p
2007]
  QL369.2.P47 2008
  571.2'94--dc22

                                        2007042877

About the Editor...

JOSÉ MANUEL PÉREZ MARTÍN was born in Madrid, Spain in 1960. In 1983, he obtained a bachelors degree in Biochemistry and Molecular Biology at the Autonomous University of Madrid (Spain) and completed his molecular biology training in 1985. In 1992, he obtained his Ph.D. degree in Molecular Biology at the Autonomous University of Madrid. Dr. Pérez Martín had a postdoctoral position from 1992 to 1996 conducting clinical trials with antitumor drugs under the sponsorship of the Bristol Myers-Squibb Pharmaceutical Research Institute (Wallingford, Connecticut, USA, and Madrid, Spain). From 1997 to 2006 he was Associate Profesor at the Faculty of Sciences of the Autonomous University of Madrid (Spain), teaching inorganic and bioinorganic chemistry. At present, he is Head of Laboratory of the Farmacia "Castilla" in Madrid (Spain). His research is focussed on the interactions between drugs and DNA or proteins and the pharmacological modulation of antitumor drugs. He is a co-author of more than 90 scientific papers including reviews and book chapters and is co-editor of the book *Metal Compounds in Cancer Chemotherapy* (Research Signpost, 2005). He is also a member of the Advisory Board of the scientific journals: *Current Medicinal Chemistry, Medicinal Chemistry, Anticancer Agents in Medicinal Chemistry*, and *Recent Patents on Anticancer Drug Discovery*.

Dedication

This book is dedicated to the memory of my late father and also to my mother, wife and son.

As an amateur guitarist, I would also like to dedicate this book to the virtuoso Spanish guitarists Fernando Sor, Dionisio Aguado, Francisco Tárrega, Andrés Segovia, Narciso Yepes and Paco de Lucia.

CONTENTS

EDITOR

José Manuel Pérez Martín
Centro de Biología Molecular
"Severo Ochoa" (CSIC-UAM)
Universidad Autónoma de Madrid
Madrid, Spain
Email: josemanuel.josema@gmail.com
Chapter 1

CONTRIBUTORS

Note: Email addresses are provided for the corresponding authors of each chapter.

Carlos Alonso
Centro de Biología Molecular
    "Severo Ochoa" (CSIC-UAM)
Universidad Autónoma de Madrid
Madrid, Spain
Chapter 1

Bijoylaxmi Banerjee
Division of Molecular Parasitology
Indian Institute of Chemical Biology
Kolkata, India
Chapter 5

Luke A. Baton
W. Harry Feinstone Department
    of Molecular Microbiology
    and Immunology
Bloomberg School of Public Health
Johns Hopkins University
Baltimore, Maryland, U.S.A.
Chapter 7

Marlene Benchimol
Laboratório de Ultraestrutura Celular
Universidade Santa Úrsula
Rio de Janeiro, Brazil
Email: marleneben@uol.com.br
Chapter 9

Josefina Castilla
Farmacia "Castilla"
Hermanos García Noblejas
Madrid, Spain
Chapter 1

Viola Denninger
Department of Biochemistry
University of Tuebingen
Tuebingen, Germany
Chapter 4

Marcel Deponte
Adolf Butenandt-Institute
    for Physiological Chemistry
Ludwig Maximilians University
Munich, Germany
Email: marcel.deponte@gmx.de
Chapter 8

Elen M. de Souza
Laboratório de Biologia Celular, DUBC
Instituto Oswaldo Cruz, FIOCRUZ
Rio de Janeiro, Brazil
Chapter 3

Silvia Díaz
Departamento de Microbiología-III
Universidad Complutense (UCM)
Madrid, Spain
Chapter 12

George Dimopoulos
W. Harry Feinstone Department
    of Molecular Microbiology
    and Immunology
Bloomberg School of Public Health
Johns Hopkins University
Baltimore, Maryland, U.S.A.
Email: gdimopou@jhsph.edu
Chapter 7

Michael Duszenko
Department of Biochemistry
University of Tuebingen
Tuebingen, Germany
Email: michael.duszenko@uni-tuebingen.de
Chapter 4

Katherine Figarella
Department of Biochemistry
University of Tuebingen
Tuebingen, Germany
Chapter 4

Miguel A. Fuertes
Centro de Biología Molecular
   "Severo Ochoa" (CSIC-UAM)
Universidad Autónoma de Madrid
Madrid, Spain
Chapter 1

Andrea Gallego
Departamento de Microbiología-III
Universidad Complutense (UCM)
Madrid, Spain
Chapter 12

Juan C. Gutiérrez
Departamento de Microbiología-III
Universidad Complutense (UCM)
Madrid, Spain
Chapter 12

Seth A. Hoffman
W. Harry Feinstone Department
   of Molecular Microbiology
   and Immunology
Bloomberg School of Public Health
Johns Hopkins University
Baltimore, Maryland, U.S.A.
Chapter 7

Hemanta K. Majumder
Division of Molecular Parasitology
Indian Institute of Chemical Biology
Kolkata, India
Email: hkmajumder@iicb.res.in
Chapter 5

Edgar Marchán
Laboratorio de Biología Molecular
Instituto de Investigaciones en
   Biomedicina y Ciencias Aplicadas
Universidad de Oriente
Cumaná, Venezuela
Chapter 6

Ana Martín González
Departamento de Microbiología-III
Universidad Complutense (UCM)
Madrid, Spain
Email: anamarti@bio.ucm.es
Chapter 12

Maribel Navarro
Laboratorio de Química Bioinorgánica
Centro de Quimica
Instituto Venezolano de Investigaciones
   Cientificas (IVIC)
Caracas, Venezuela
Email: mnavarro@ivic.ve
Chapter 6

Paul A. Nguewa
Centro de Biología Molecular
   "Severo Ochoa" (CSIC-UAM)
Universidad Autónoma de Madrid
Madrid, Spain
Chapter 1

Alberto Roseto
Laboratoire Génie Enzymatique
   et Cellulaire, UMR CNRS 6022
Université de Technologie de Compiègne
Compiègne, France
Email: alberto.roseto@utc.fr
Chapter 2

Claude-Olivier Sarde
Département de Génie Biologique
Université de Technologie de Compiègne
Compiègne, France
Chapter 2

María Segovia
Departamento de Ecología
Universidad de Málaga
Malaga, Spain
Email: segovia@uma.es
Chapter 11

Nilkantha Sen
Division of Molecular Parasitology
Indian Institute of Chemical Biology
Kolkata, India
Chapter 5

Maria de Nazaré C. Soeiro
Laboratorio Biologia Celular, DUBC
Instituto Oswaldo Cruz, FIOCRUZ
Rio de Janeiro, Brazil
Email: soeiro@ioc.fiocruz.br
Chapter 3

Kevin S.W. Tan
Laboratory of Molecular and Cellular
    Parasitology
Department of Microbiology
Yong Loo Lin School of Medicine
National University of Singapore
Singapore
Email: mictank@nus.edu.sg
Chapter 10

Néstor L. Uzcátegui
Department of Biochemistry
University of Tuebingen
Tuebingen, Germany
Chapter 4

Gonzalo Visbal
Laboratorio de Síntesis Orgánica y
    Productos Naturales
Centro de Química
Instituto Venezolano de Investigaciones
    Científicas (IVIC)
Caracas, Venezuela
Chapter 6

Emma Warr
W. Harry Feinstone Department
    of Molecular Microbiology
    and Immunology
Bloomberg School of Public Health
Johns Hopkins University
Baltimore, Maryland, U.S.A.
Chapter 7

Susan Welburn
Department of Biochemistry
University of Tuebingen
Tuebingen, Germany
Chapter 4

PREFACE

Under the name of programmed cell death (PCD) are included diverse molecular mechanisms of cell suicide which play an essential role in the development of multicellular organisms. The best known PCD mechanism in multicellular organisms is called apoptosis. However, recent studies indicate that PCD is also present in protozoa and unicellular eukaryotes. The twelve chapters of this book give the reader a comprehensive update of the progress in the understanding of the mechanisms of PCD in protozoa. The chapters have been written by experts in this field of research and are arranged following an evolutionary point of view starting with PCD in protists and ending with PCD in ciliated protozoa.

Chapter 1 is an overview of the current knowledge about PCD in protozoa using the example of kinetoplastid parasites as unicellular eukaryotic (mitochondriate) organisms that inherited PCD from ancestral prokaryotic protozoa. Chapter 2 deals with the intriguing fact that in amitochondriate organisms where only hydrogenosomes and mitosomes subsist as mitochondrial relics, recent findings show that PCD also occurs. This exciting discovery is presented here in the light of sequencing in various species as well as recent findings about mitochondrial derivates and ancestral viruses, contributing to a better understanding of the life tree as well as to the future discovery of new molecules of interest. In Chapter 3 the phenomenon of apoptosis, one of the types of PCD, is reviewed in three vector-borne trypanosomatids (*Trypanosoma cruzi*, *Trypanosoma brucei*, and *Leishmania spp*) responsible for diseases of great medical and veterinary importance. Chapter 4 summarizes the most obvious findings regarding programmed cell death in African trypanosomes. In Chapter 5, the use of topoisomerase inhibitors as tools to disentangle the molecular mechanisms of PCD in the kinetoplastid parasite *Leishmania* is presented. Chapter 6 reports on the uses of metal complexes as leishmanicidal drugs, especially those having leishmanicidal activity which could be linked to their interaction with the parasitic DNA. In addition, PCD-inducing drugs used clinically against Leishmaniasis and those currently in the experimental and evaluation phases are reviewed. Chapter 7 deals with the fact that in recent years there has been an increasing awareness of the role of PCD in the malaria parasite's infection of its vertebrate host and mosquito vector. In fact, a significant body of research now indicates that PCD of both vertebrate host and mosquito vector cells plays an important, if still incompletely understood, role during infection with this parasite. The understanding of this role may have medical applications in the treatment of malaria. In Chapter 8 the hypothesis that some stages of malaria parasites (*Plasmodium*) are able to undergo a form of PCD is supported by available data. Moreover, this chapter presents the current knowledge on *Plasmodium* metacaspases; these putative proteases are the most promising candidates that might be essential for the execution of PCD in malaria parasites. Chapter 9 is mainly devoted to the important study of PCD in *Trichomonas foetus*, a cattle parasite, and *Trichomonas vaginalis*, a human parasite. *Trichomonads* do not possess mitochondria but harbor another type of membrane-bounded organelle, an unusual anaerobic energy-producing organelle known as the hydrogenosome. Studies of cell death in trichomonads are under way in order to establish whether the hydrogenosome could represent an alternative to mitochondria and whether these organisms possess all caspase activities and which conditions lead *Trichomonads* to cell death. The presence of a cell death program in *Trichomonads* suggests the existence either of a dependent or independent caspase-like execution pathway in such organisms. In Chapter 10, PCD in the enteric protozoon parasite *Blastocystis hominis* is described. Chapter 11 presents PCD in *Dinoflagellates* which are protists ecologically important as components of phytoplankton. The acquisition of PCD genes in *Dinoflagellates* goes back to ancient times where endosymbiotic events took place. Chapter 12 reports on the fascinating phenomenon of programmed nuclear death (PND) in ciliated protozoa. The main objective of this PND is to remove the old macronucleus while a new recombinant vegetative nucleus develops in each conjugating cell. The mechanism of PND is still not elucidated, but we know that it involves caspase-like proteins, an intense acid phosphatase activity and an autophagic process.

Finally, I would like to thank Landes Bioscience and all the authors participating in this book for the patience they have had with the Editor during the two years that have passed since the original project of this book was conceived.

José Manuel Pérez Martín, Ph.D.
Madrid, Spain

Acknowledgements

The Editor is very grateful to Ron Landes who suggested the idea of publishing a book of this ilk. Many thanks to Cynthia Conomos, Bonnelle Martin and Celeste Carlton from Landes Bioscience for their superb and hard work in the publication of the book.

CHAPTER 1

Programmed Cell Death in Protozoa:
An Evolutionary Point of View.
The Example of Kinetoplastid Parasites

Miguel A. Fuertes, Paul A. Nguewa, Josefina Castilla, Carlos Alonso
and José Manuel Pérez Martín*

Abstract

Programmed cell death (PCD) is a molecular event which plays an essential role in the development of multicellular organisms. However, recent studies indicate that PCD is a mechanism also present in protozoa and unicellular eukaryotes. For instance, it has been recently proposed that some *Trypanosomatid* parasites have a PCD mechanism descendant from an ancient life form that has actually evolved. Thus, two hypotheses may explain the existence of PCD in protozoa such as *Trypanosomatids*. First, PCD could simply be a process without a defined function inherited through cell evolution, which is triggered in response to diverse stimuli and stress conditions. Alternatively, PCD might be used by *Trypanosomatids* as a control mechanism to maximize their biological fitness.

Introduction

Figure 1 shows that diverse forms of programmed cell death (PCD) have been recently described in at least nine species of protozoa, whose phylogenic divergence is believed to range from around two to one billion years ago.[1] Some of these PCD forms have been reported in the kinetoplastid parasites of the genuses *Trypanosoma* and *Leishmania* that are believed to be amongst the earliest diverging eukaryotes. These kinetoplastid parasites are the agents responsible for trypanosomiasis and leishmaniasis, tropical illnesses that suffer approximately 30 million people around the world.[2] It is interesting to know that the cell death phenotype of the kinetoplastid parasites shares several features with apoptosis which is the mechanism of cell death shown by multicellular organisms. In fact, PCD in kinetoplastid parasites includes cytoplasmic blebbing and vacuolization, chromatin condensation and DNA fragmentation. These findings indicate that PCD may have evolved together with the endosymbiotic incorporation of aerobic bacteria (the precursors of mitochondria) into ancestral unicellular eukaryotes.[3] Hence, two hypothesis may account for the existence of PCD in single-celled organisms such as *Trypanosomatids*. On the one hand, PCD in *Trypanosomatids* could simply be a remnant process of eukaryotic cell evolution without a particular function, which is induced in response to diverse stimuli (for example, serum removal, oxidants such as H_2O_2 and chemotherapeutic agents).[4] On the other hand, PCD could have a defined biological role for *Trypanosomatids* as a way to maximise the biological fitness of these parasites facilitating their adaptation to a digenic life cycle (mammalians-insect-mammalian). Therefore, of particular interest is the question of the origin and nature of PCD in protozoa as well as its important role in the

*Corresponding Author: José Manuel Pérez Martín—Centro de Biología Molecular "Severo Ochoa" (CSIC-UAM), Facultad de Ciencias, Universidad Autónoma de Madrid, Cantoblanco, 28049-Madrid, Spain. Email: josemanuel.josema@gmail.com

Programmed Cell Death in Protozoa, edited by José Manuel Pérez Martín.
©2008 Landes Bioscience and Springer Science+Business Media.

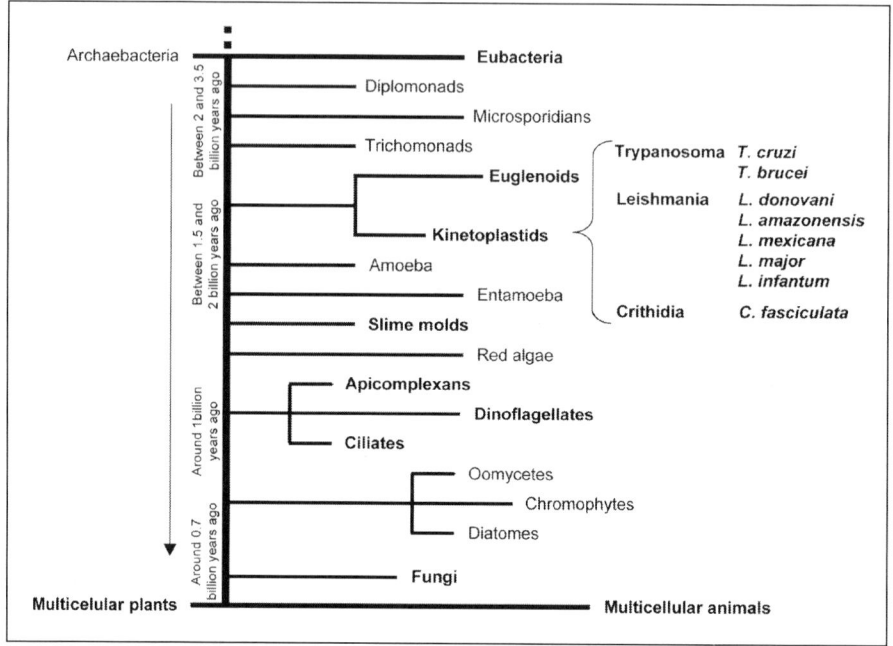

Figure 1. Programmed cell death (PCD) during evolution. Within the phylogenic tree, PCD has been identified in various members (in bold) of the tree including *Trypanosomatids*.

regulation of the complex interactions between unicellular and multicellular organisms which allow, for instance, the establishment and persistence of stable host/parasite interactions. Last but not least, the extent of overlapping between effectors and regulators of PCD among *Trypanosomatids* and mammalian hosts PCD pathways may determine whether or not the pathways leading to cell suicide in these parasites may be pharmacologically exploited as a parasite control strategy.

The Example of PCD in Kinetoplastids as an Heritage of Evolution

Much evidence accumulated over the last years indicates that PCD in protozoa comes from an ancestral death machinery.[5] As mentioned above, *Kinetoplatid* parasites are eukaryotic protozoa that belong to one of the most ancient diverging branches of the eukaryotes phylogenic tree.[6,7] These single-celled organisms are reported amongst the first mitochondrial eukaryotes and contain only one giant mitochondrion, called the kinetoplast.[8]

First of all, we shall discuss about the evolutionary implications of mitochondrial acquisition in life/death regulation. Thus, we should take into account that programmed cell death may be particularly useful when cells are interacting. For instance, in trypanosomes PCD may control communication between unicellular and multicellular organisms, which allows the establishment of a stable host-parasite relationship.[9,10] But, how and when did protozoa choose genes allowing cell suicide? Firstly, it has been hypothesized that a common PCD mechanism arose before the evolution of multicellularity.[11] Secondly, the hypothesis that the eukaryote cell is a symbiont that emerged from the fusion of different bacteria species suggests that PCD may have evolved from a resolution of conflict between heterogeneous genomes within a cell, which subsequently led to an enforced cooperation.[1] Hence, it is likely that PCD may have evolved together with the endosymbiotic incorporation of aerobic bacteria (the precursors of mitochondria) into ancestral unicellular eukaryotes.[3] In fact, it is gaining recognition the hypothesis that eukaryotic cells are descendants of ancient anaerobic organisms that survived, in a world that had become rich in oxygen, by engulfing aerobic bacteria, keeping them in symbiosis for the sake of their ability to consume atmospheric oxygen and produce energy as ATP. So, mitochondria would originate from Krebs-cycle-containing eubacteria (promitochondria) invading a fermentative anaerobe.[3,12] Then, during eukaryotic evolution, most of the genetic information contained in the promitochondrial genome was incorporated into the nuclear genome.[13] This symbiotic process is believed to have helped the ancestral eukaryotic cells to utilize the oxidative metabolism of the symbiotic bacteria to gain energy as the eukaryotes adapted to the change from an anaerobic environment to an increasing oxygen-rich surrounding.[14] As a result, over the next few hundred million years, the symbiotic bacteria lost most of their essential genes except some of those required for oxidative respiration and ATP synthesis, giving some of the genes to the host nucleus whose proteins would still target the symbiont for activity. This process is hypothesized to have directed the ancestral bacterial symbionts to become obligate endosymbionts and a de facto eukaryotic organelle known as mitochondrion.[15-17] Additional evolutionary divergence led to different eukaryotic organisms having different numbers of mitochondria.

It has been recently reported that some present day prokaryotes liberate redox proteins that induce apoptosis in eukaryotic cells through stabilization of p53 protein.[14] Thus, it has been proposed that the parents of the present day prokaryotes released redox proteins to kill the ancestors of the eukaryotes. Subsequently, during the evolution of mitochondria as obligate endosymbionts, mitochondrial ancestors offered some useful functions to their hosts, which in turn provide the formers with physical protection and essential nutrients. In this way, the mitochondria adapted their original "killing" functions to programme their host cell death. As it is known for mitochondria in multicellular organisms, kinetoplasts might not only be powerhouses for generation of cellular energy but also organelles, which play a major role in inducing PCD of the parasite through the release of redox proteins.[14]

Although apoptosis or developmentally regulated programmed cell death is probably only present in multicellular organisms,[18,19] it seems that some forms of PCD may have evolved at the same time as did endosymbiosis. Hence, it is of crucial importance to know the reasons whereby the basic mechanisms of PCD were established in protozoa, particularly in a number of single-celled eukaryotes, such as *Trypanosoma cruzi*,[9] *Trypanosoma brucei rhodesiense*[10] and *Leishmania amazonensis*,[20] *L. donovani*[21,22] and *L. major*.[8] The understanding of those reasons would also help to explain PCD phenomena described in fungi and plants.[3]

The Hypothesis of PCD in *Trypanosomatids* as an Adaptative Mechanism and a Defence Strategy

As we have discussed in the previous section, PCD in *Trypanosomatids* may be a remnant process of cellular evolution without a specific function. However, it is gaining recognition the hypothesis that considers PCD in *Trypanosomatids* as a pathway to maximise their biological fitness. In this context, PCD in *Trypanosomatids* might serve as a molecular mechanism of adaptation and defence against the host.

One of the proposed functions of the PCD pathway in unicellular protozoa is to control cell population, as is the case of multicellular organisms.[11,23-25] On the other hand, it is well-known that programmed cell death takes place during the digenic life cycle of *Trypanosomatids*. In fact, it has been hypothesized that PCD might act during the digenic cycle of *Leishmania*. So, some individuals may be programmed to infect and suffer "terminal differentiation", while others are ancillary to the former and still others are present to maintain a certain population density for the infection.[20,26] It is interesting to point out that procyclic trypanosomes displaying morphological features of apoptosis have been found in the midgut of flies.[27] It is known that for their survival within the sand fly gut (or within the hosts), *Trypanosomatids* must have a thorough control to restrict individuals; otherwise death of the insect vector (or the macrophages) may occur prematurely. In this scenario, PCD could be triggered to accomplish this required control because of the competition of the parasites for the limited resources in the sand fly gut (or within the hosts). For example, in *Trypanosoma brucei*, PCD could be an important event in the tsetse fly, where careful parasite population size control operates.[28] Since parasites and flies compete for the aminoacid proline, the maintenance of such an equilibrium, in which parasite multiplication is compensated by parasite death, can be mutually helpful.[29]

Recently, it has been proposed that mammalian host cells can trigger apoptosis rather than necrosis of some *Trypanosomatids*.[30] Thus, in the three severe human disease agents, *Trypanosoma cruzi*, *Trypanosoma brucei* and *Leishmania amazonensis*,[31] PCD could be regulated by signals from their multicellular hosts, such as temperature and lectins, as well as by components of the host immune system including complement proteins and cytokines.[32] Massive death of kinetoplastid parasites has also been postulated as an evolutionary process in which the non-adapted parasites die by apoptosis, probably, for the maintenance of an immunological silent state of the host during the infection process.[30] On the other hand, it has been reported that phagocytosis of apoptotic cells favours the intracellular growth of *Trypanosoma cruzi*.[33] So, PCD may be a suitable strategy of defence, which permits the parasitic infection to "go-ahead". This mechanism of cell death may also result beneficial for other *kinetoplastid* parasites allowing not only their adaptation to the external condition and their development but also protecting them against possible damages from the hosts.

PCD might also serve as a cell sorting mechanism to select specific parasitic forms (for example the metacyclic form from the procyclic one). This mechanism of cell sorting would ensure the selection of the infectious parasitic form, since uninfectious forms no longer contribute to the perpetuation of parasite life cycle and however may compete with the differentiated *Trypanosomatids* for available nutrients.

Altogether the above-mentioned data suggest that in *Trypanosomatids* the regulation of PCD might allow a careful coupling of appropriate cell differentiation and cell survival. In this context, PCD induction through complex parasite-mammalian host interactions would be in agreement with the central role that has been attributed to apoptosis in multicellular organisms, namely, the thorough control of cell differentiation, the matching of cell numbers to their environment and the defence against genetic damage and infections, leading to the elimination of abnormal and infected cells.

Future Prospects

When dealing with programmed cell death in protozoa we need to keep two things in mind for the future. First, we need to be aware of the differences, as well as the similarities with multicellular organisms. Hence, convergence and divergence of primitive characteristics of PCD in protozoa and mammalians cells might be useful fingerprints to establish evolutive relationships. In fact, studies in *Trypanosomatids* suggest the existence of effector and regulator molecules as caspase-like, poly(ADP-ribose) polymerase-like (PARP-like), GSH-like (trypanothione) that exhibit similar activities to the observed in mammalian PCD phenomena (Fig. 2). Moreover, *Trypanosomatids* have a phylogenetically mitochondrial-originated protein called metacaspase,[34,35] which might have evolved to the today-known caspases of multicellular organisms. Second, from a pharmacological point of view, we have to take into account that depending on whether infectious single-celled organisms share some or all of the effectors and regulators common to multicellular apoptosis or have evolved their own divergent pathways, we will have to use different therapeutic approaches to induce specific killing by PCD.[11] For example, it has been found that the lipophilic drug o-naphthoquinone β-lapachone inhibits a PARP-like enzyme isolated from the *Trypanosomatid Crithidia fasciculate*.[36] Because PARP enzymes are involved in recognition of DNA damage and induction of cell death, inhibition of PARP activity might be used in the future to increase the antiparasitic effects of DNA-binding drugs such as pentamidine.

Acknowledgements

This work was supported by Spanish Comisión Interministerial de Ciencia y Tecnología (Grant SAF 2004-03111) and by Fondo de Investigaciones Sanitarias (Grant C03/04). We also

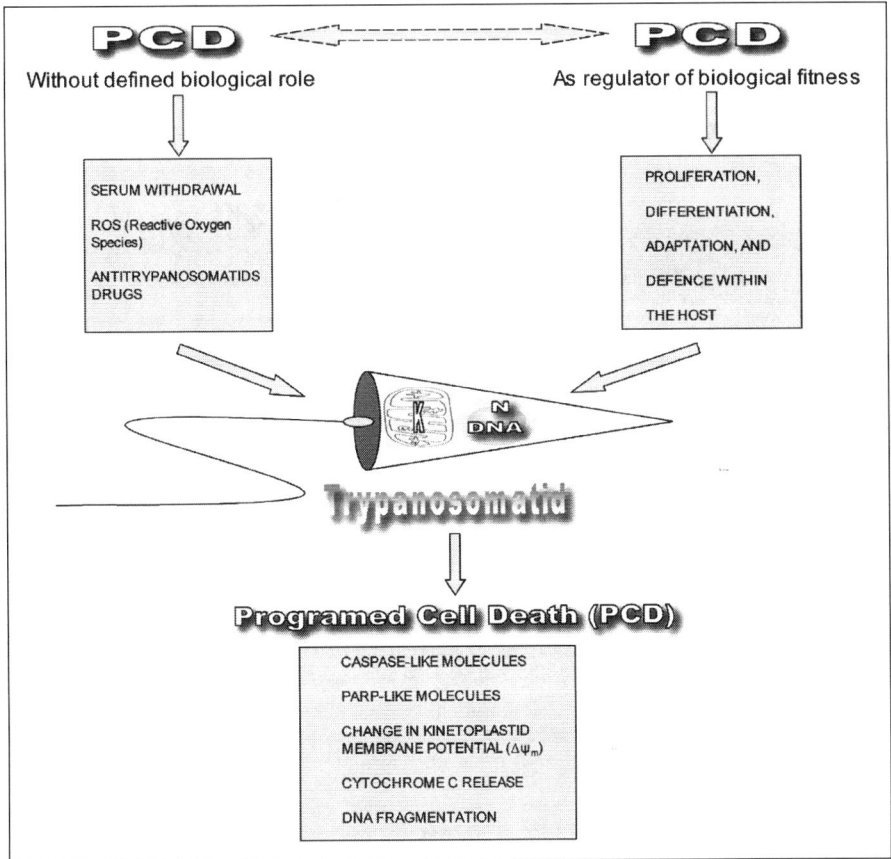

Figure 2. Two hypotheses for the existence of PCD in *Trypanosomatids*. PCD in *Trypanosomatids* may be a remnant process of cell evolution without a defined function, which is activated in response to diverse stimuli (serum deprivation, ROS, cytotoxic drugs, etc.). Alternatively, PCD may serve as a way to maximise the biological fitness of the parasites by controlling proliferation, differentiation, adaptation and defence within the host. In this latter case, PCD will be also triggered in response to stress conditions (discontinuous arrow). K = kinetoplastid, N = nucleus.

thank sponsorship by European COST Action D20/003/00 and by a CDTI grant to laboratorios CBF-Leti. An institutional grant from Fundación Ramón Areces is also acknowledged.

References

1. Ameisen JC. The origin of programmed cell death. Science 1996; 272:1278-1279.
2. World Health Organization The World Health report: life in the 21st century. A vision for all. Report of the Director-General, WHO, Geneva, 1998:44-51.
3. Kroemer G. Mitochondrial implication apoptosis. Towards an endosymbiont hypothesis of apoptosis evolution. Cell Death Differ 1997; 4:443-456.
4. Zangger H, Mottram JC, Fasel N et al. Cell death in Leishmania by stress and differentiation: programmed cell death or necrosis? Cell Death Differ 2002; 9:1126-1139.
5. Fraser A, James C. Fermenting debate: do yeast undergo apoptosis? Trends Cell Biol 1998; 8:219-221.
6. Sogin ML. Early evolution and the origin of eukaryotes. Curr Op Gen Dev 1991; 1:457-463.
7. Doolittle RF, Feng DF, Tsang S et al. Determining divergence times of the major kingdoms of living organisms with a protein clock. Science 1996; 271:470-477.

8. Arnoult D, Akarid K, Grodet A et al. On the evolution of programmed cell death: apoptosis of the unicellular eukaryote Leishmania major involves cysteine proteinases activation and mitochondrion permeabilization. Cell Death Diff 2002; 9:65-81.
9. Ameisen JC, Idziorek T, Billaut-Mulot O et al. Apoptosis in a unicellular eukaryote (Trypanosoma cruzi): implications for the evolutionary origin and role of programmed cell death in the control of cell proliferation, differentiation and survival. Cell Death Differ 1995; 2:285-300.
10. Welburn SC, Dale C, Ellis D et al. Apoptosis in procyclic T. brucei. rhodesiense in vitro. Cell Death Differ 1996; 3:229-236.
11. Welburn SC, Barcinski MA, Williams GT. Programmed cell death in Trypanosomatids. Parasitol Today 1997; 13:22-26.
12. Margulis L. Symbiotic theory of the origin of eukaryotic organelles. Jenning DH, Leed DL, eds. Symbiosis. Symposium 29: Society for Experimental Biology. Cambridge: Cambridge University Press, 1975:21-38.
13. Gray MW. Origin and evolution of mitochondrial DNA. Annu Rev Biochem 1989; 5:25-50.
14. Punj V, Chakrabarty AM. Redox proteins in mammalian cell death: an evolutionarily conserved function in mitochondria and prokaryotes. Cell Microbiol 2003; 5:225-231.
15. Blackstone NW. A units-of-evolution perspective on the endosymbiont theory of the origin of the mitochondrion. Evolution 1995; 49:785-796.
16. Margulis L. Archaeal-eubacterial mergers in the origin of Eukarya: phylogenetic classification of life. Proc Natl Acad Sci USA 1996; 93:1071-1076.
17. Doolittle W. Eukaryote origins: a paradigm gets shifty. Nature 1998; 392:15-16.
18. Raff M. Social control of cell survival and cell death. Nature 1994; 365:397-400.
19. Fuertesa MA, Castillab J, Alonsoa C, Pérez JM. Cisplatin biochemical mechanism of action: from cyto-toxicity to induction of cell death through interconnections between apoptotic and necrotic pathways. Curr Med Chem 2003; 10:257-266.
20. Moreira ME, Del Portillo HA, Milder RV et al. Heat shock induction of apoptosis in promastigotes of the unicellular organism Leishmania amazonensis. J Cell Physiol 1996; 167:305-313.
21. Das M, Mukherjee SB, Shaha C. Hydrogen peroxide induces apoptosis-like death in Leishmania donovani promastigotes. J Cell Sci 2001; 114:2461–2469.
22. Lee N, Bertholet S, Debrabant A et al. Programmed Cell Death in the unicelular protozoan parasite Leishmania. Cell Death Differ 2002; 9:53-64.
23. Lymbery AJ, Hobbs RP, Thompson RC et al. Building bridges and controlling parasites. Int J Parasitol 1997; 27:1119-1120.
24. Anderson RM. Complex dynamic behaviours in the interaction between parasite population and the host's immune system. Int J Parasitol 1998; 28:551-66.
25. Barcinski MA, DosReis GA. Apoptosis in parasites and parasite-induced apoptosis in the host immune system: a new approach to parasitic diseases. Braz J Med Biol Res 1999; 32:395-401.
26. Vickerman K. Developmental cycles and biology of pathogenic trypanosomes. Br Med Bull 1985; 41:105-114.
27. Welburn SC, Maudlin I, Ellis DS. Rate of trypanosome killing by lectins in midguts of different species and strains of Glossina. Med Vet Entomol 1989; 3:77-82.
28. DosReis GA, Barcinski MA. Apoptosis and parasitism: from the parasite to the host immune response. Adv Parasitol 2001; 49:133-161.
29. Welburn SC, Maudlin I. Tsetse-trypanosome interactions: rites of passage. Parasitol Today 1999; 15:399-403.
30. Piacenza L, Peluffo G, Radi R. L-Arginine-dependent suppression of apoptosis in Trypanosoma cruzi: Contribution of the nitric oxide and polyamine pathways. Proc Natl Acad Sci USA 2001; 98:7301-7306.
31. Maslov DA, Simpson L. Evolution of parasitism in kinetoplastid protozoa. Parasitol Today 1995; 11:30-32.
32. Ameisen JC. On the origin, evolution and nature of programmed cell death: a timeline of four billion years. Cell Death Diff 2002; 9:367-393.
33. Freire-de-Lima CG, Nascimento DO, Soares MB et al. Uptake of apoptotic cells drives the growth of a pathogenic trypanosome in macrophages. Nature 2000; 403:199-203.
34. Koonin EV, Aravind L. Origin and evolution of eukaryotic apoptosis: the bacterial connection. Cell Death Differ 2002; 9:394-404.
35. Mottram JC, Helms MJ, Coombs GH, Sajid M. Clan CD cysteine peptidases of parasitic protozoa. Trends Parasitol 2003; 19:182-187.
36. Villamil SF, Podestá D, Molina Portela MD, Stoppani A. Characterization of poly(ADP-ribose) polymerase from Crithidia fasciculata: enzyme inhibition by β-lapachone. Mol Biochem Parasitol 2001; 115:249-256.

CHAPTER 2

Programmed Cell Death in Protists without Mitochondria:
The Missing Link

Claude-Olivier Sarde and Alberto Roseto*

Programmed cell death (PCD), a fundamental process that can be triggered in all cells, was supposed until recently solely centred on the mitochondrion. However, in amitochondriate organisms where only hydrogenosomes and mitosomes subsist as mitochondria relics, recent findings show that PCD still occurs. This exciting discovery is presented here in the light of the development of sequencing project in various different species as well as recent findings about mitochondrial derivates and ancestral viruses, contributing to a better understanding of the life tree as well as to the future discovery of new molecules of interest.

Life Origin

Since the emergence of the cell theory at the end of the XIXth century, cells are known to be the fundamental units of life. Either isolated or grouped in complex organisms, they form all the living beings found in every place reached so far by man, from atmosphere to the deepest parts of soils and oceans.[1,2] Even in the most unfavourable physicochemical conditions encountered (absolute anaerobiosis, highly acidic medium, high salt concentration, elevated radiations pressure), cells still present a common face.[3,4] They are all formed of the same type of macromolecules, among which nucleic acids play a central role. They all are able to auto-replicate, and all of them use a similar genetic code to translate their genome into building proteins and enzymes. They also form "molecular nanomachines", complex suprastructures, which can regroup up to a hundred different macromolecules (e.g., ribosomes, spliceosomes, etc).[5,6] Of course, discrepancies can be noted here and there.[7-9] But a kind of unity is so obvious that it gives even to distant species a kinship aspect. Inherited from the long lasting pre-eminence of the Darwinist point of view, it is thus tempting to consider the evolutionary process as a tree, where all living beings have evolved from the same ancestral cell. In this evolutive view, mutations are thought to affect the progeny and thus give birth to various branches. As time passes by, these branches are naturally selected to finally generate cell diversity.[10-13] Based on this concept, the living world was divided during the past decade into three domains: two of prokaryotic (Archaea and Bacteria) and one eukaryotic (Eucarya) nature, all however derived from a common root called LUCA, the last universal common ancestor.[14-16] This approach was strengthened by molecular sequence analysis of translational machinery components such as small sub-unit ribosomal RNA or translational elongation factor genes. In this view, archea, which were supposed to be involved in the acquisition of an alpha-protobacteria able to produce its own energy, were regarded more or less as the primitive essence of eukaryotes.[13,14,17]

*Corresponding Author: Alberto Roseto—Laboratoire Génie Enzymatique et Cellulaire, UMR CNRS 6022, Université de Technologie de Compiègne, B.P. 20529, 60205 Compiègne cedex, France. Email: alberto.roseto@utc.fr

Programmed Cell Death in Protozoa, edited by José Manuel Pérez Martín.
©2008 Landes Bioscience and Springer Science+Business Media.

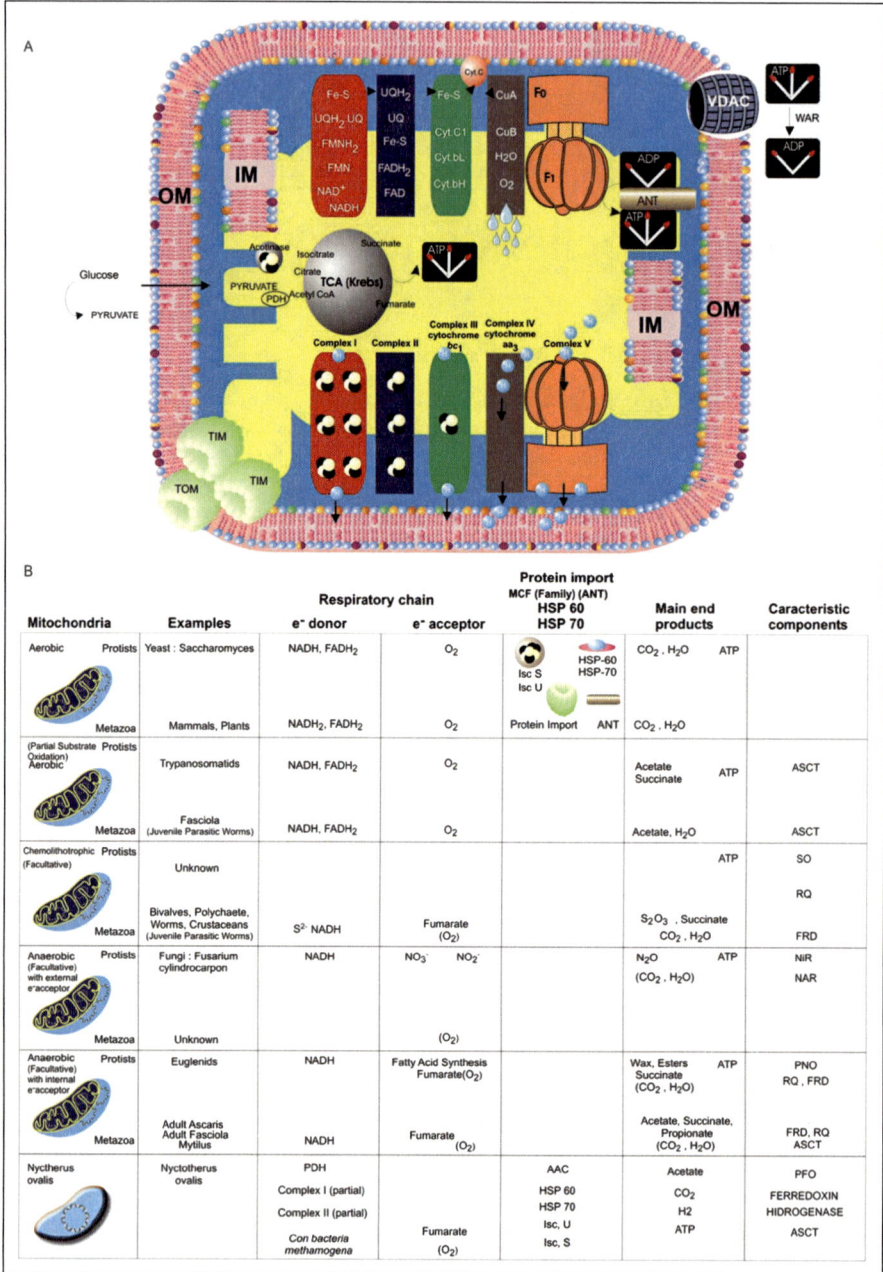

Figure 1. The main characteristics of aerobic and anaerobic mitochondria. Aerobic and anaerobic mitochondria are regarded as the major features of eukaryotic cells. The typical aerobic
mitochondria (encountered e.g., in mammals) requires O_2 for its function and produces ATP.
The pyruvate dehydrogenase is responsible for the oxidative decarboxylation of pyruvate into
acetyl CoA, which is further completely oxidized to CO_2 in the Krebs cycle (tricarboxilic acid
cycle, TCA). The figure legend is continued on the next page.

Figure 1, continued. But the maximum amount of energy is produced through mitochondrial oxidative phosporylation (OXPHOS) and culminates in complex V which imports H and produces ATP. A) Diagram showing the basic relationships between OXPHOS and TCA and ATP energy production relatively to the number of the Isc (Fe:S cluster). Complex I (red) has six Isc (black and yellow circles), complex II (blue) has three Isc and complex III (green) has one Rieske center. In TCA, the cis-acontinase, which converts citrate to isocitrate, contains a 4Fe-4S center. Fe-S are absent in Complex IV (brown) and V (purple) of the OXPHOS. Contact between the outer membrane (OM) and the inner membrane (IM) in classical mitochondria. The OM are exclusively the VDA (voltage dependent anion channel), which allows diffusion of metabolites and ions across the OM. The IM is impermeable, allowing the maintenance of a transmembrane potential. The IM has transport molecules as ANT (adenine nucleotide transporter) responsible for the passage of ADP:ATP. The majority of mitochondrial proteins are encoded by nuclear DNA and imported from cytoplasm into mitochondria by translocator of the outer membrane (TOM) and of the inner membrane (TIM) complexes (globally named IMP, imported membrane proteins). B) An important number of eukaryotes have a reduced sensitivity to the lack of oxygen. Some can survive by simple fermentation (e.g., Yeast), an exclusive cytosolic process which ends up with lactate or ethanol. But the mitochondrial synthesis of ATP can also be very diverse. In organisms that live during all or part of their life cycle in low oxygen (microaerophilia) or in oxygen–free (anoxia) conditions, the anaerobic mitochondria produces ATP with the H$^+$ pumping electron transport, and do not need O_2. Acceptors other than O_2 are shown in the respiratory chain (acceptors could be external to the cell, e.g., NO_3 or produced inside of the cell, e.g., fumarate). The presence of Fe:S centers and of vital chaperonines 60 and 70 are signaled. The characteristic enzymes in each group are indicated. (Modified from ref. 32).

But bacteria may use other ways than the simplistic mutation-selection events in order to remodel their DNA. Indeed, the engulfment of neighbouring cells and the stable acquisition of their DNA content offers tremendous possibilities of genomic plasticity. (This DNA may either maintain independently within internal organelles through endosymbiosis as it supposed for mitochondria and chloroplasts or be directly incorporated in the genome of the recipient cell by recombination) During this acquisition phase, the genetic union of highly divergent cell lineages recomposes the genetic background even more than a sexual mating would do.[18] On the other hand, conjugative nucleic acid transfers are among the most efficient ways for bacteria to reorganize their genetic content, easily acquiring genes and operons from other species in the same way as grafts allow the remodelling of fruit trees.[17-19] Of course, nobody can tell how often these processes have occurred in life's history (although we know the second is still very frequent and can affect very distant cell types as it is the case e.g., between Agrobacterium and plant cells,[19-21] but it seems obvious that they were essential in the early time, providing sources of some convergent evolution which is not allowed by the single Darwinist process.

The observations above were reinforced when observing the overall evolution of nanomachines or rare changes in complex molecular features where an ancestral state can be inferred, such as e.g., the separation or fusion of the dihydrofolate reductase (DHFR) and thymidylate synthase (TS)[22] genes and the presence or not of a type II myosin gene in eukaryotes.[23] Such results have lead to modify the eukaryotic branching which previously distinguished true eukaryotes from Archezoas, [The Archezoa contained the diplomonads (e.g., *Giardia*), parabasalids (e.g., *Trichomonas*), archamoebae (e.g., *Entamoeba*), and the microsporidia (e.g., *Trachipleistophora*)], that were characterized by their lack of mitochondria and regarded as primitive.[14] The new branching is now located between the unikonts (regrouping animals, fungi and amoebozoa) and bikonts (including plants, algae and most protozoa). This also forces to reconsider the "primitive" state of amitonchrials organisms, since it appears that mitochondria was "lost" instead of "not acquired" as previously thought.[13] Indeed, newer findings have demonstrated that organisms devoid of mitochondria not only have mitochondrial-derived proteins that are now encoded directly by their genome, but also, doubled membrane-bounded organelles with remnants of mitochondrial function (see below) (Fig. 1A).[24-26]

Until recently, viruses, the most abundant "organisms" of the Earth, were considered as clearly distinct entities, being not truly "alive" since they do not obey the fundamental rules of cells. Indeed, viruses cannot replicate their nucleic acids or translate their proteins alone, borrowing most of the necessary components from the host cell. Even if not "alive", viruses exhibit in their architecture and propagation mode the same relative homogeneity that is observed in living beings.[27-29] This homogeneity also suggests a long-range evolutionary relationship between viruses affecting the three main branches of the tree: archea, bacteria and eukaria. In the attempt to explain this, the putative existence of a common ancestor that would have preceded the divergence of the three domains of life three billion years ago is in favour with scientists. The exact nature of this ancestor is still unknown, but the recent discovery of a giant virus growing in amoebae, Mimivirus, allows some speculations.[30] Therefore, if the "tree" notion that describes the relationship between organisms is probably true for late appearing species, in which gene transfer processes are less frequent, it is now called into question for primitive ancestors and their progeny. From this point of view, the early time of life should rather be seen as a "mangrove" with numerous intricated and fused roots that have exchanged their genetic content almost ad libitum.[20,31]

The Mitochondrial Role

Mitochondria, together with nucleus, Golgi apparatus, and endoplasmic reticulum are generally considered to be the basic characteristics of all eukaryotic cells. Harbouring their own DNA circular genome that varies in size and copy number between species, they play a central role in energy production through the import from cytoplasm and conversion of pyruvate into acetyl-coenzyme A (CoA) and its further oxidation in CO_2 during the citric acid cycle. This oxidation is associated with the production of NADH and $FADH_2$ reducing power molecules which are in turn reoxidized by the respiratory chain located in the mitochondrial inner membrane. Oxidation reactions are coupled with the translocation of protons across the inner membrane which generates the protomotrice force necessary to drive ATP synthesis by the ATP synthase as well as to translocate metabolites across the membrane. Beside this central process, mitochondria also ensure the degradation of fatty acids via beta-oxidation, parts of the urea cycle, Fe-S complex assembly, and the regulation of calcium homeostasis. Moreover, citric acid cycle delivers metabolic intermediates for amino acid and nucleotide biosynthesis[32-34] (Fig. 1A,B). Last but not least, mitochondria are now considered as the intracellular integrators of programmed cell death (PCD) in which they play a central role[35] (see below).

From Morphology and Energy Production to Classification

Mitochondria can also exhibit a highly modified aspect with a considerable reduction of structural organization and functions.[24] One can distinguish different types of organisms that lack typical mitochondria according to the different variants observed. Type I totally lacks a mitochondrion but the existence of this kind of organism is only putative since no member has been discovered so far. Type II possesses a mitochondrial derivate called hydrogenosome keeping the energy production function, but lacking a genome (except in *Nyctotherus ovalis*).(Fig. 2A,B) Type III includes mitochondrial relics named mitosomes, which lack both energy function and genome (except in *Blastocystis hominis* which harbour DNA). All mitonchondrion derivates are self-replicating and segregate into daughter cell after cell division. All have also kept their original double membrane.[36-38] (Fig. 3A)

Hydrogenosomes are hydrogen-producing organelles found in a variety of anaerobic unicellular eukaryotes. The pathway leading to hydrogen production is composed of three enzymes: pyruvate: ferredoxin oxidoreductase complex (PFO), ferredoxin, and hydrogenase, an enzyme which is not present in mitochondria. Pyruvate produced by glycolysis is decarboxylated by PFO into acetyl-CoA (in aerobic eukaryotes acetyl-CoA is produced exclusively inside mitochondria). This results in the concomitant production of energy in the form of ATP by substrate-level phosphorylation using the citric acid cycle enzyme succinyl: CoA-transferase. Simultaneously, H_2 is released by hydrogenase[39-41] (Fig. 2A)

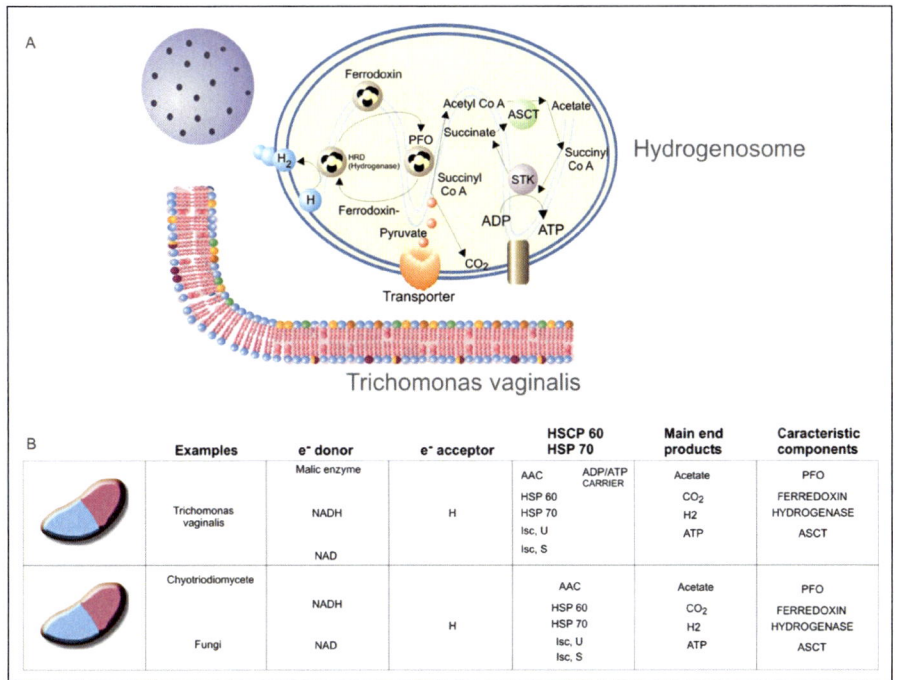

Figure 2. The hydrogenosome. A) In various taxa of the life tree, another form of anaerobic production of ATP is present. Hydrogenosomes are mitochondria-derived organelles (MDO) producing H_2 and ATP and found in amitochondriate protists including parabasalides (T. vaginalis, chytridiomycetes fungi, ciliates, ameboflagellates). When compared with aerobic and anaerobic mitochondria, their main characteristics are: no membrane-associated electron transport chain; no DNA (so far, the only known exception is the organelle found in *Nyctotherus ovalis*, that we and others consider as a mitochondrion producing H and ATP); decarboxylation of pyruvate in acetate and CO_2 with concomitant production of ATP; production of H_2. The pathway of this production requires the unique and specific hydrogenosome hydrogenase (HDR), the pyruvate:ferrodoxin oxidoreductase (PFO) and the Ferredoxin (Fd). All three enzymes posess a Fe-S cluster center. The chaperonines (CNP) 60 and 70 are located inside the hydrogenosome. A homolog of the mitochondrial ADP:ATP translocator (AAC) is also present. B) Like aerobic and anaerobic mitochondria, hydogenosomes possess the key component of the vital Fe-S cluster. In this respect, examples shown here conform to the expected pattern for mitochondrial relics.

Mitosomes are mitochondrion-derived organelles that have lost their energy metabolism. Indeed, very few mitosomal proteins have been detected through direct observation, among which hsp,60 hsp70 and Fe-S complex (IscS, IscU). First discovered in *Entamoeba Histolytica*,[42,43] similar remnants were found in *Trachipleistophora hominis*,[44] *Cryptosporidium parvum*,[45] *Encephalitozoon cuniculi*,[46,47] *Giardia intestinalis*[48] and *Blastocystis hominis*.[37] (Fig. 3)

How PCD Was Born

As mentionned above, the endosymbiotic theory places the mitochondrion inside all primitive cells located within the early roots of the life tree(Fig. 4A). It is also now well accepted that mitochondrion plays a central part in all the different forms of programmed cell death (the most common form being apoptosis) identified in all metazoans models investigated so far (see Fig. 3)

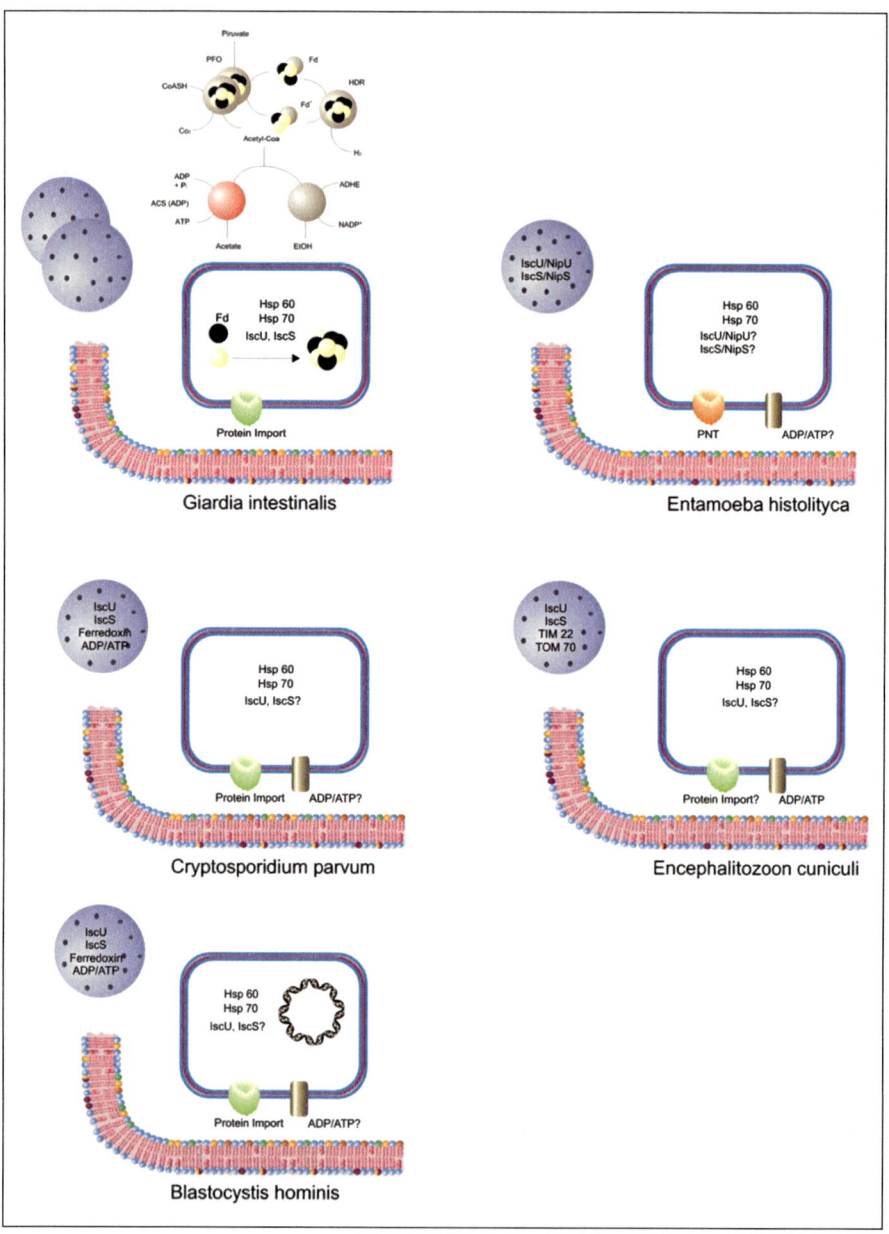

Figure 3. Mitosomes. Mitosomes (also named crypton or relictual mitochondria) are mitochondria-derived organelles (MDO) that have lost the essential components of energy metabolism. They do not produce energy of any kind. They do not harbour DNA (the only exception is the organelle of *B. hominis*). As in hydrogenosomes, the presence inside the mitosome or in the cytoplasm of gene products such as HSP60 and 70, both vital for the assembly of Fe-S center, show the mitochondrial origin of the organelle. Fe-S cluster is also present in other organelles (by direct or indirect demonstration) and is actually the only remnant of mitochondria in mitosome. The vital function of this complex may be the only function of the organelle actually indispensable for life.

In mammals, eleven ways of dying have been described until now, ten of which appear to be genetically programmed: Apoptosis, Anoikis, Caspase Independent Apoptosis, Autophagy, Wallerian degeneration; Excitotoxicity, Erythropoiesis, Platelets (PLT), Cornification of Skin, Lens, and Necrosis (non programmed). Some other reported death forms are only variants of the same process (oncosis vs necrosis and anoikis vs cell detachment induced apoptosis).[49,50]

Molecular mechanisms as well as biochemical and morphological observation about PCD are summarized as apoptosis, a term introduced in 1972 by Kerr, Wyllie and Currie.[(i)51,52] In contrast to necrosis, which is a pathological condition resulting of severe cellular damage, apoptosis is a naturally occurring, controlled, and strictly regulated process that can be induced by one or several internal or external signals. When dying cells are examined microscopically, the main visible intracellular event is chromatin processing. During nonnecrotic death, the chromatin of dying cells condenses and simultaneously undergoes fragmentation of its DNA content into nucleosomal fragments, a phenomenon known as apoptotic body formation. Protein cleavage also results in morphological modifications, all contributing to the hallmark changes in apoptotic cells, including nuclear membrane breakdown and externalization of phosphatidylserine (PS) residues.

Different pathways can lead to apoptosis: two intrinsic pathways and a extrinsic pathway.[53,54] The extrinsic pathway involves a death receptor protein (e.g., Fas/CD95) and an adaptor protein (e.g., FADD, Fas-associated death domain protein), which in turn interacts with the cysteine aspartate protease pro-caspase-8. Activation of caspase-8 results in the activation of the caspases cascade, which includes the executioner caspases-3,-6 and -7 and indirectly triggers the activation of caspase-3 together with the mitochondrial release of cytochrome c. In fact, pro-apoptotic proteins resulting from activation of death pathways move either from the cytoplasm [e.g., B-cell leukaemia/lymphoma 2 (Bcl-2) interacting mediator of cell death (Bim), Bcl-2-antagonist of cell death (Bad), BH3- interacting domain death agonist (Bid), stress-activated protein kinase (SAPK), c-JUN N-terminal protein kinase (JNK)] or from the nucleus [e.g., thyroid hormone receptor 3 (TR3), tumor suppressor protein p53] to mitochondrial membranes where they interact with receptors such as Bcl-2, B-cell lymphoma-x long (Bcl-xl), Bcl-2-associated X protein (Bax), or permeability transition pore complex (PTPC). This leads to the loss of integrity of the mitochondrial membrane, thus generating a mitochondrial membrane permeabilization and the release of apoptogenic factors. Other 'secondary' pro-apoptotic molecules are also produced by mitochondria including calcium ions, ceramides, ganglioside GD3, sphingosines, palmitate, reactive oxygen species (ROS) and nitric oxide.

Two intrinsic parallel pathways have been characterized in which mitochondrion plays a central role: the caspase-dependent and the caspase-independent pathways. In the caspase-dependent intrinsic pathway, when mitochondria receive one or more of these death signals, cytochrome c is released in the cytosol where it interacts with apoptotic protease activation factor 1(Apaf-1) to form the apoptosome. Apoptosome interacts with pro-caspase-9 leading to the formation of the caspase-9 activation complex that stimulates the caspases cascade. The mitochondrial permeabilisation allows the release of second mitochondria-derived activator of caspases/direct IAP-binding protein with low pI (Smac/DIABLO) and High temperature resistant A2/OMI[(ii)] (HtrA2/OMI)[55,56] that neutralise the inhibitory apoptosis proteins (IAPs)], triggering in turn the caspases pathway. In the caspase-independent intrinsic pathway, endonuclease G, and apoptosis inducing factor (AIF) are released from the mitochondrial space to the cytoplasmic compartment.[53,58-60] (Fig. 5A).

The entire process, from the initial inductive signal to the destruction of the cell, can take hours or even days, but the early events (mitochondrial permeabilisation) occur in 10 minutes. It

---

i "The use of the word apoptosis can be found for the first time between the end of the 5th century BC and the early 4th century BC, in chapter 35 of *Mochlicon*—Hippocrates' treatise on the reduction of dislocations. In this chapter, the word apoptosis is used to describe gangrene resulting from treatments of fractures with bandages"(*Nicolas André, The Lancet*).

ii OMI is not an acronym, but a real name given "in memoriam" of her beloved dog by L. Faccio (personal communication).

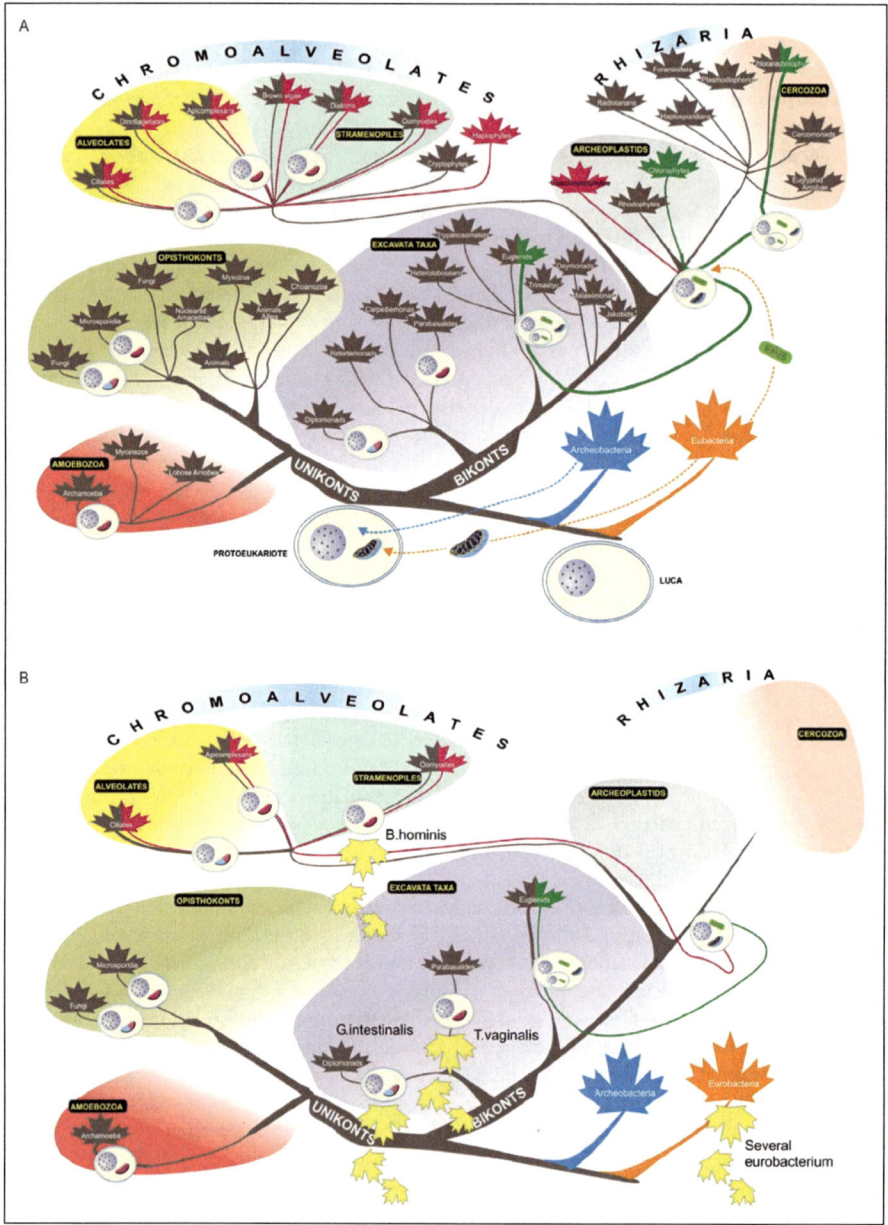

Figure 4. The tree of life. A consensus life tree can be designed from the integration of recent works of different laboratories together with existing data but the true starting point as well as branching point of diverses subgroups remain uncertain (modified from ref 13). A) Most important branches in the three kingdoms, hypothetic origin of LUCA and pre- LUCA period (see text). The eukaryotes phylum reflect the current theory about its subdivision between unikonts and bikonts into two big sub-branches. Unikonts are composed of Opisthokonts and Amoebozoa. Bikonts regroup Chromoalveolates, Archaeplastids, Excavata and Rhizaria. Lineages containing species, where the presence of hydrogenosomes (red-blue oval) or of mitosomes (red oval) has been reported, are indicated. Figure legend continued on next page.

Figure 4, continued. Both are present in different branches of both unikonts and bikonts, supporting the idea that the lack of mitochondria is a state neither rare nor primitive.The position of amitochondriate organisms in the life tree, we and others, assume that the endosymbiosis of an alpha protobacteria by a post-LUCA organism is at the origin of the Eukaryotes "big bang". The secondary endosymbiosis of plastides (green oval) or of other unicellular organism (small cell inside another cytoplasm) are indicated by red and green branches respectively. B)PCD forms are present in eubacteria as well as in amitochondriate organisms (from either unikonts or bikonts branch) located in the deepest roots of the life tree. The absence of classical mitochondria as well as of caspases (key element of PCD in all other mitochondriate organisms) is not "an impediment". The discovery of new ancestral molecules participating in PCD process in all living beings (including amitochondriate ones) may give the life tree a stronger coherence.

is strongly suspected that once these events proceed to the point of executioner caspase cascade activation without constraint, the death of the cell may be inevitable.[61]

Recently, new partners have been added to the long list of proteins able to inducee apoptosis. Indeed, when a T cytotoxic lymphocyte (TC cell) reacts against a tumoral or infected cell, it does it through the binding between TCR and foreign peptide-associated HLA complex. Consequently to this binding, TC cell secretes perforin that generates a hole in the target cell allowing in turn the entrance of molecules from the Granzyme family (namely GzmA and GzmB).[62,63] It appears that mitochondrion is involved in the apoptotic process triggered by GzmB since apoptosome formation and caspases activation have been reported. On the contrary, GzmA seems to be involved in a mitochondrion-independent apoptotic pathway, by targeting the endoplasmic reticulum-associated SET complex, which contains the GzmA-activated DNase/Non metastatic 23 H1 (GAAD/NM23-H1), its inhibitor Suvar-Enhancer of Zeste-Trithorax (SET), Abasic endonuclease 1 (Ape1) and the 3' prime repair exonuclease (TREX1). Single-stranded DNA damage is initiated when activated NM23-H1 nick DNA after GzmA cleaves its inhibitor SET.[64,65] (Fig. 5A,B)

PCD in Bacteria and Mitochondriate Unicellular Organisms

Most PCD molecular studies were initially done in multicellular organisms. The abundance of sequencing project in all kingdoms and the generalization of the bioinformatic tool allows now to investigate the presence of the principal molecules implicated in the PCD pathways in various living beings. Surprisingly, these molecules are scarcely present in unicellular organisms. However the morphological features resembling the classical PCD observed in metazoars can be observed in prokarya as well as in unicellular mitochondriate or amitochondriate organisms.[50,66,67]

Growing experimental evidences have revealed the existence of programmed cell death (PCD) systems in bacteria, but real observations are still scarce. Among reported cases, one can cite the regulable suicide module mazEF which is activated by several antibiotics in *Eschericha coli* and several other pathogenic bacteria.[68,69] In cyanobacteria such as *Trichodesmium* sp. IMS101, death process is associated with an increase of caspase-like activity and can be triggered in response to phosphorous and iron starvation, increased light irradiance and oxidative stress.[70] In *Anabaena spp.*, an exposure to moderate concentration of univalent-cation salts is sufficient to initiate cell death associated with a notable increase in nonspecific protease activity.[71]

In silico studies, together with the rapidly growing number of publications on multiple apoptotic markers in unicellular organisms support the theory that PCD is widespread among them. Species explored so far for the existence of PCD regroup phytoplankton unicellular members,[72] *Dictyostelium discoideum*,[73,74] *Tetrahymena thermophila*,[75,76] *Leishmania major*,[77] *Leishmania infatum*,[78] *Leishmania donovani*,[79] *Trypanosoma cruzi*,[80,81] *Trypanosome brucei rhodesiense*,[82] *Aspergillus fumigatus*,[83] *Saccharomyces cerevisiae*,[84,85] *Plasmodium berguei*,[86] *Plasmodium falciparum*.[87]

Phytoplankton evolved in the Archaean oceans more than 2.8 billion years ago and is of crucial importance in regulating aquatic food webs, biogeochemical cycles and the Earth's climate. Until recently, phytoplankton was considered immortal unless killed or eaten by predators.[72] However, over the past decade, it has become clear that these organisms can either be infected by viruses (e.g., *Emiliania huxleyi* Prymnesiophyte) or undergo self-destruction

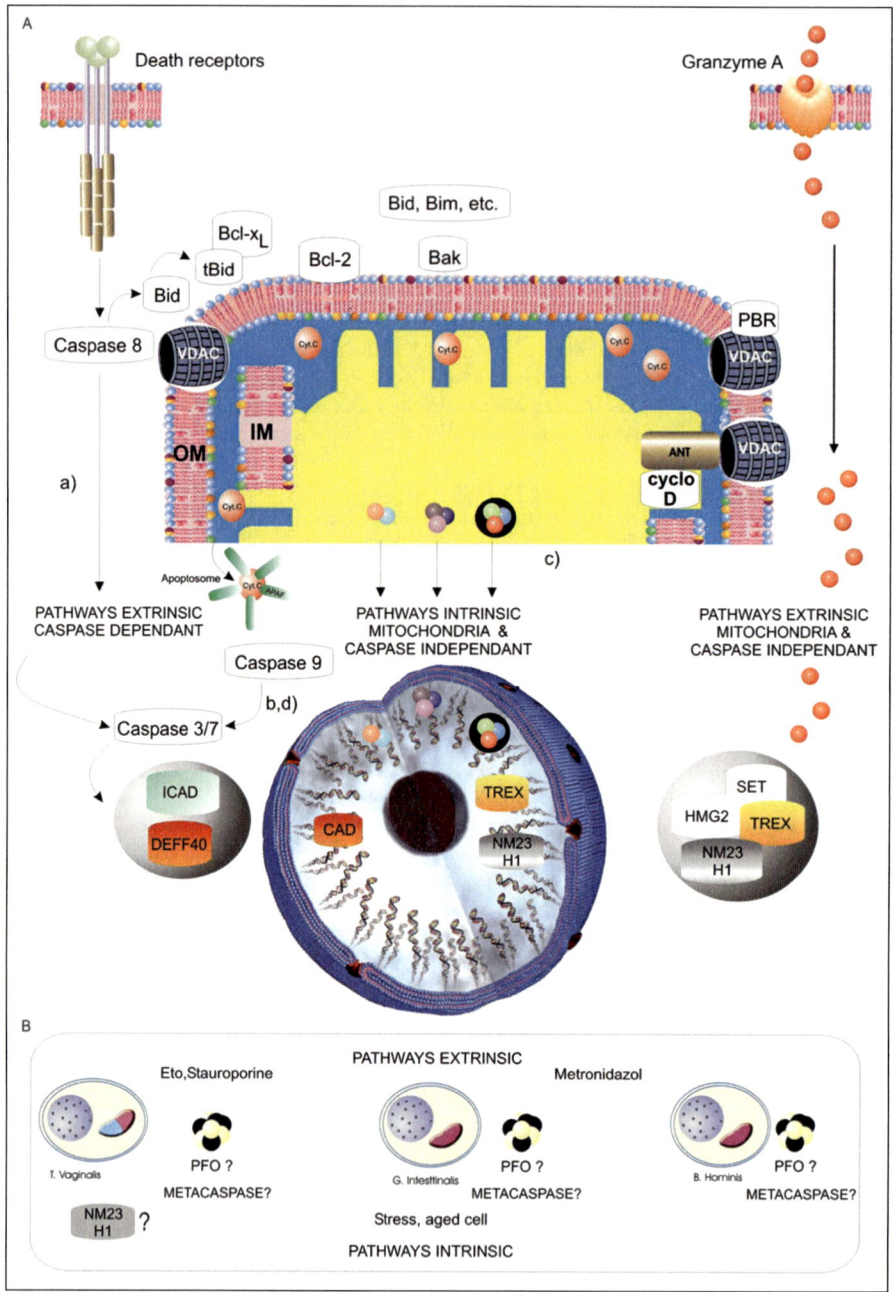

Figure 5. The pathways of apoptosis. A) PCD in aerobic metazoars follows three main pathways: 1) extrensic pathway: caspase-dependent (exclusive, whithout the participation of the mitochondria) (a) or activating the mitochondria (more frequent) (b). In general, both work simultaneously. 2) intrinsic pathway when the apoptotic stimulus comes from inside the cell. This pathway is always mitochondria-dependent. It can be exclusive whithout caspases involvement (c) or, mitochondria- and caspases-dependent (d). Both can occur together. Figure legend continued on next page.

Figure 5, continued. 3) The third pathways is caspase- and mitochondria-independent. In this pathway, Granzyme A directly affects the cytoplasmatic proteic complex composed, so far, by SET, TREX, HMG2, pp32 and NM23-H1. TREX and NM23 nucleases are both implicated in DNA fragmentation (see text). B) The apoptosis in mitochondriate and amitochondriate unicellular organisms. In unicellular organisms with aerobic or anaerobic mitochondria and harbouring metacaspases (caspases do not exist in protozoars), the different forms of PCD are conditionned by the participation of the mitochondria and /or the different metacaspases present (see text). In amitochondria protists that only possess hydrogenosomes or mitosomes, few if any molecular data have been obtained showing the implication of new proteins. If one consideres that *B. hominis* is an amitochondriate organism of type III, the participation of PFO in its PCD may be suspected. In *T. vaginalis*, *G. intestinalis* and *E. histolytica*, the implication of PFO can also be suspected when those organisms are in presence of metronidazole. On the other hand, comparative bioinformatic investigations in *T. vaginalis* and *G. intestinalis* have shown the presence of homologs of the NM23 gene family, the participation of which into PCD process is under way (COS, AR personal communication).

mechanism analogous to multicellular organisms PCD in response to environmental stress (such as cell age, nutrient deprivation, intense light, excessive salt concentrations or oxidative stress.[70,72] The diatoms *Ditylum brightwellii*[88] and *Thalassiosira weissflogii*,[89] the dinoflagellates *Peridinium gatunese*[90] and *Amphidinium carterae*,[91] the chlorophyte *Dunaliella tertiolecta*,[92] the coccolithophore *Emiliania huxleyi*[72] are representative exemples of species in which PCD has been reported so far.

In the single-celled soil-living amoeba *Dictyostelium discoideum*, PCD can be induced as a monolayer autophagic vacuolar cell death in a caspase-like-independent manner with several apoptosis-like features. The presence of a potential homolog of mammalian AIF (DdAIF) has been reported. The autophagy process has been demonstrated to be partially dependent of the atg1 gene.[73,74] The ciliated protozoan *Tetrahymena thermophila* has a unique apoptosis-like nuclear death called programmed nuclear death (PND) that occurs during conjugation and in which caspase-8- and -9-like were shown to be implicated. An incomplete degradation of nuclear DNA can be observed. The participation of mitochondrial endonuclease G has also been demonstrated. The implication of both caspases and mitochondrial pathways in *Tetrahymena thermophila* PCD is highly probable.[75,76] In *Leishmania major*, staurosporine induces a PCD form with several hallmarks of apoptosis.[77] Similar morphologic results can also be reproduced in *L. infantum* when incubated at temperatures above 38°C[78] or in *L. donovani* upon induction by H_2O_2.[79] *L. major* mitochondria released cytochrome c and cysteine proteinases with nuclear pro-apoptotic activity.[77] In *Trypanosoma cruzi*, a PCD process can be induced by drugs, nutritional and oxidative stress, heat shock and exposure to human serum factor.[80,81] The presence of Bcl homologues and metacaspases has been reported.[80] Procyclic *Trypanosoma brucei rhodesiense* exhibits a cell death mechanism which can be activated by an external signal, the lectin ConA, in vitro.[82] Fungi like *Aspergillus fumigatus* can undergo autophagic- or apoptotic-type PCD on exposure to antifungal agents, developmental signals, and stress factors. Filamentous fungi can also exhibit a form of cell death called heterokaryon incompatibility (HI) triggered by the fusion between two genetically incompatible individuals. Using a systematic computational approach, more than a hundred PCD related genes potentially involved in HI have been recently identified.[83] In the baker yeast *Saccharomyces cerevisiae*, several studies involving mutations and mutant rescue indicate that apoptosis-like programs are indeed present. On the other hand, treatments with e.g., H_2O_2 and acetic acid induce yeast cell death in a process accompanied by typical apoptotic changes. Yeast analogues of some components of the apoptotic cascade have been described, i.e., a caspase, Omi and apoptosis-inducing factor.[84,85] A whole range of apoptotic markers has been described for *Plasmodium spp*. mainly in the mosquito midgut stages of *P. berghei* both in vivo and in vitro.[86] In *P. falciparum*, where in vivo data are scarser, the existence of PCD is mainly supported by in silico elements.[87]

PCD in Amitochondriate Unicellular Organisms

As discussed above, one of the most crucial steps during the evolution of eukarya toward the explosion of the metazoan world was certainly the acquisition of the mitochondrion. Moreover, all protists cited in which a PCD process has been described harbour this organelle in their cytoplasm. In amitochondriate organisms investigated until now, the absence of this central organelle may be partially compensated by remnants of mitochondria (either hydrogenosomes or mitosomes). Consequently, no homologues of the molecules described in metazoans to participate to PCD could be detected. However, a PCD process has been described to occur in *Trichomonas vaginalis*,[94] *Giardia intestinalis*[95] and *Blastocystis hominis*.[96] (Fig. 4A,B).

Trichomonas vaginalis

T. vaginalis is the most common nonviral sexually transmitted disease, infecting more than a hundred million people annually. Trichomoniasis is associated with many perinatal complications, male and female genito-urinary tract infections, and an increased incidence of HIV transmission. The preferred drug of treatment is metronidazole, a 5-nitroimidazole derivative. Resistance results from loss of hydrogenosomal proteins such as PFO, ferredoxin, hydrogenase, and malic enzyme.[98]

The paraxostylar and paracostal granules found in *Trichomonas vaginalis* are typical hydrogenosomes, that link this parabasalid to Type II amitochondriate protists. They are DNA-free double-membrane-bound organelles producing H_2, a metabolic end-product highly unusual among eukaryotes. The presence of enzymes not found in mitochondria (e.g., pyruvate: ferrodoxin oxidoreductase and hydrogenase) and the absence of a number of mitochondrial functions, (e.g., oxidative phosphorylation) are the main features of this unique organelle. It has been suggested that mitochondrion and hydrogenosome share a common ancestral origin. This idea is supported by the recent identification of Hmp31, a homologue of ADP/ATP carrier (ANT), identified in *Trichomonas* hydrogenosomal membrane.[40,41]

T. vaginalis morphological aspects after treatment with proapoptotic drugs ETO, DOXO and STS show some features resembling apoptosis. Indeed, "apoptotic-like" conformations including nuclear fragmentation, chromatin condensation and apoptotic-like bodies formation (the latest close resembling apoptotic bodies) can be observed. When the classical pan-caspase pathway inhibitor z-VAD.fmk is used in combination with pro-apoptotic drugs ETO, DOXO and STS, it does not abolish cell death when *T. vaginalis* is exposed to ETO and DOXO, but a weak response can observed after exposure to STS. This result does not exclude the presence of more ancestral caspase-like protein such as metacaspases and paracaspases. A positive reactivity to anti-caspase-3 monoclonal antibody could also be observed. The positive TUNEL staining in *T. vaginalis* treated with ETO and STS shows that *T. vaginalis* DNA also undergoes fragmentation, but without laddering or high-molecular weight banding. The same feature can be observed in "naturally" dead cells (issued from stress-induced natural-occuring death after at least 72 h of culture), inferring that a caspase-like pathway (or more ancient equivalent proteases such as those described in *C. elegans*) may exist in amitochondrial protists. In addition, the exposure of PS on the outer leaflet of the plasma membrane, another event that is one of the hallmarks of apoptosis, can be induced in the *T. vaginalis* with the STS pro-apoptotic drug.[98]

Blastocystis hominis

Blastocystis hominis is a human intestinal parasite implicated in several intestinal diseases like diarrhea and irritable bowel syndrome. *B. hominis* is a single-celled organism but has multiple nuclei and a variable number (from one to four up to a few hundred) of mitochondria. Since *B. hominis* is a strict anaerobe, its mitochondria have been submitted to various studies which have revealed the absence of many characteristic mitochondrial features. *B. hominis* mitochondrial relic contains DNA but no mitochondrial electron transport chain. It was recently suggested that in the absence of any evidence of energy generation, the *B. hominis* mitochondria should be considered as mitosomes despite the presence of a genome.

In this organism, a PCD form can be triggered by external induction. Indeed, when treated with the surface reactive cytotoxic mAb 1D5 (an antibody raised against B. hominis isolate B) or with the metronidazole drug, key morphological and biochemical features of apoptosis can be observed, such as e.g., nuclear condensation, fragmentation of DNA and reduced cytoplasmic volume.[99,100]

Giardia intestinalis

Giardia intestinalis (also known as *Giardia lamblia*) is a flagellated protist of the diplomonad group that commonly causes intestinal malabsorption and diarrhea throughout the world. All known free-living and parasitic diplomonads share some unique structural features, including the absence of mitochondria, peroxisomes, and a recognizable Golgi apparatus.[101,102] As shown by scanning and transmission microscopy, trophozoites of *G. intestinalis* have two similar-looking nuclei. Both are transcriptionally active. When *G. intestinalis* cells were treated with pro-apoptotic drugs, or stressed by nutriment depletion, a form of cell death with some but not all features of canonical PCD could be observed, very similar to the one described in *T. vaginalis*. Interestingly, *G. intestinalis* treated with metronidazole presented the same morphological features.[95] When epifluorescence microscopy was performed, the various aspects of chromatin condensation described in PCD for unicellular and multicellular organisms harbouring mitochondria were present. When tested, an effective action of caspases activators and inhibitors was ascertain, suggesting the presence of caspase-like proteases involved in PCD in this protist. Until now, however, *G. intestinalis* sequence analyses performed at protein and DNA level have always failed to detect an equivalent for every known PCD-involved molecules, this including AIF and caspases-related compounds. The observation that metronidazole might induce programmed cell death is interesting since this drug needs a low redox potential to be activated. In trichomonads *G. intestinalis* and *B. hominis*, this is achieved by the action of PFO, which in trichomonads resides in the hydrogenosomes. This might suggest the presence of PFO in the mitosomes of *B. hominis* and *G. intestinalis*.[24]

Perspectives

PCD is a normal physiological process that is known to be triggered by internal or external stimuli. Initially discovered in metazoans, it is now described in all organisms from all parts of the phylogenic tree investigated so far. Indeed, during the last years, various forms of cell death morphologically and molecularly resembling metazoans PCD have been identified in unicellular mitochondriate and energy producing protists. Thus, one should now add the different forms of PCD observed in prokarya, mitochondriate and amitochondriate protozoan to the 11 ones first described in mammals. However, such a demultiplication of PCD forms is sometimes hard to admit for purists, for whom some distant variants should almost be regarded as more or less degenerated. Indeed when looking at the early bases of the branches of the phylogenic tree, PCD process seems barely to keep a wireframe of common morphological features. Sometimes, it also unambiguously harbours unexpected variations which prevent it to fit, even roughly, with the "canonical" forms observed in metazoans.

This "golden apple of discord" is mainly due to the fact that PCD morphological characterisation is essentially performed through the observation of features that are the hallmark of apoptosis, the prototype of PCD in higher eukaryotes. Apoptosis, the first process to be described and from far the more studied throughout the scientific world, is now widely used as the "bona fide" reference in order to evaluate what should or should not be called PCD. If the need of such a reference is not questionable, the conclusion drawn from the comparison should be treated with more caution. Indeed, if one single apoptotic marker, taken alone, is no proof of PCD, it seems however justified to classify some of the processes observed in unicellular organisms as true PCD forms (distinct from metazoans PCD) when several apoptotic markers can be reproducibly shown to occur. Moreover, our view mainly centered on mammals and man should not occult the fact that organisms in which the referential (namely apoptosis) was chosen, are actually neither the most abundant nor the most diverse species represented on earth surface. From this point of view, the unicellular world is far richer than the pluricellular one will ever be. Keeping this is mind, the observations made for PCD in amitochondrial protists take of course a very different meaning.

From what we know now of the origin of life, mitochondrion appears more and more as a kind of Gordian knot, perhaps not totally untractable, but which, alas, for the time being, cannot be solved by a bold stroke. This knot forms the trunk between the intricate roots deeply soaked in the primitive mangrove and the evolved branches of the living world (Fig. 4). It obviously contains the elements necessary to the comprehension of when and how the pre-LUCA or LUCA acquired its nucleus and mitochondria. But the existence of PCD in amitochondriate organisms is not only a subject for philosophic discussion upon the origin of life and mitochondrial evolution. It is also a highly interesting source of fundamental and practical approaches to more concrete problems. As classically described in metazoans, the two molecular pathways (caspases-dependent and caspases-independent) described so far to be involved in PCD processes need mitochondria involvement. It is only recently that the identification of Granzyme A and of its role in PCD through the induction of NM23-H1 has cast the basis of a new pathway that appeared until now dissociated from the already described ones. This new finding certainly constitutes a different way of approach in the attempts to control cell death in mammals and finally in man. In this purpose, the comprehension of PCD in amitochondriate organisms will probably give an interesting scope angle on a phenomenon in which mitochondrial participation, if it exists, will be anyway regarded as marginal.

When analysing in mitochondrial derivatives (the fact that hydrogenosomes and mitosomes are indeed evolved forms of mitochondria is now widely accepted) the remnants of what became, as time was passing by, the essence of eukaryote, the only thing we can observe is the presence of Fe-S Complex. This complex, which is built through the action of about ten molecules, directly participates in the function of over a hundred proteins of vital importance implicated in the replication, translation and primary metabolism of all living cells. But it has never been shown to be directly implicated in death pathways (its only implication appears to be indirect through the formation of oxygen-reactive species). Therefore, the final lesson that can be drawn from amitochondriate organisms studies is also very probably that the true key elements, the most ancient way for a cell to control its way to die may not lie in the mitochondrion as initially thought. It may be possible that the recent discovery of Mimivirus and the probable subsequent integration of the virus world into the living beings can bring new insights into this fascinating attempt to understand how life and death were simultaneously born.

References

1. Kasting JF, Siefert JL. Life and the evolution of earth's atmosphere. Science 2002; 296:1066-1068.
2. Newman DK, Banfield JF. Geomicrobiology: how molecular-scale interactions underpin biogeochemical systems. Science 2002; 296:1071-1077.
3. Carroll SB. Chance and necessity: the evolution of morphological complexity and diversity. Nature 2001; 409:1102-1109.
4. Nisbet EG, SleepNH. The habitat and nature of early life. Nature 2001; 409:1083-1091.
5. Alberts B. The Cell as a Collection Overview of Protein Machines: Preparing the Next Generation of Molecular Biologists. Cell 1998; 92:291-294.
6. Cavalier-Smith T, Brasier M, Embley TM. Introduction: how and when did microbes change the world? Phil Trans R Soc B 2006; 361:845-850.
7. Baldauf SL.The Deep Roots of Eukaryotes. Science 2003; 300:1703-1706.
8. Klobutcher LA, Farabaugh PJ. Shifty ciliates: frequent programmed translational frameshifting in Euplotids. Cell 2002; 111:763-766.
9. Moreno Diaz de la Espina S, Alverca E, Cuadrado A et al. Organization of the genome and gene expression in a nuclear environment lacking histones and nucleosomes: the amazing dinoflagellates. Eur J Cell Biol 2005; 284:137-149.
10. Schopf JW. Fossil evidence of Archaean life. Phil Trans R Soc B 2006; 361:869–885.
11. Zimmer C. How and Where Did Life on Earth Arise? Science 2005; 309:89.
12. Rees DC, Howard JB. The Interface Between the Biological and Inorganic Worlds: Iron-Sulfur Metalloclusters. Science 2003; 300:929-930.
13. Embley TM, Martin W. Eukaryotic evolution, changes and challenges. Nature 2006; 440:623-630.

14. Cavalier-Smith T-Cell evolution and Earth history: stasis and revolution. Phil Trans R Soc B 2006; 361:969-1006.

15. Martin W, Martin Embley T. Early evolution comes full circle. Nature 2004; 431:134-136.

16. Forterre P. The origin of DNA genomes and DNA replication proteins. Curr Opin Microbiol 2002; 5:525-532.

17. Woose CR, Kandler O, Wheelis ML. Towards a natural system of organisms: Proposal for the domains Archaea, Bacteria, and Eucarya. Proc Nat Acad Sci USA 1990; 87:4576-4579.

18. Dyall SD, Brown MT, Johnson PJ. Ancient invasions: from endosymbionts to organelles. Science 2004; 304:253-257.

19. Rivera MC, Lake JA. The ring of life provides evidence for a genome fusion origin of eukaryotes. Nature 2004; 431:152-155.

20. Forterre P. The origin of viruses and their possible roles in major evolutionary transitions. Virus Res 2006; 117:5-16.

21. Genetello C, Van Larebeke N, Holsters M et al. Ti plasmids of Agrobacterium as conjugative plasmids. Nature 1977; 26:561-563.

22. Philippe H, Lopez P, Brinkmann et al. Early- branching or fast- evolving eukaryotes? An answer based on slowly evolving positions. Proc R Soc Lond B 2000; 267:1213-1221.

23. Richards TA, Cavalier-Smith T. Myosin domain evolution and the primary divergence of eukaryotes. Nature 2005; 436:1113-1118.

24. van der Giezen M, Tovar J, Clark CG. Mitochondrion-derived organelles in protists and fungi. Int Rev Cytol 2005; 244:175-225.

25. Embley TM, Hirt RP. Early branching eukaryotes? Curr Opin Genet Dev 1998; 8:624-629.

26. Tovar J, Fischer A, Graham Clark C. The mitosome, a novel organelle related to mitochondria in the amitochondrial parasite Entamoeba histolytica. Mol Microb 1999; 32:1013-1021.

27. Liu J, Glazko G, Mushegian A. Protein repertoire of double-stranded DNA bacteriophages. Virus Res 2006; 117:68-80.

28. Burnett RM. More barrels from the viral tree of life. Proc Nat Acad Sci USA 2006; 103:3-4.

29. Bamford DH. Evolution of Viral Structure. Theor Pop Biol 2002; 61:461-470.

30. Suzan-Monti M, La Scola B, Raoult D. Genomic and evolutionary aspects of Mimivirus. Virus Res 2006; 117:145-155.

31. Iyer LM, Balaji S, Koonin EV et al. Evolutionary genomics of nucleo-cytoplasmic large DNA viruses. Virus Res 2006; 117:156-184.

32. Tielens AG, Rotte C, van Hellemon JJ et al. Mitochondria as we don't know them. Trends Biochem Sci 2002; 27:564-572.

33. Pfanner N, Geissler A. Versatility of the mitochondrial protein import machinery. Nat Rev Mol Cell Biol 2001; 2:339-349.

34. Gray MW, Lang F, Burger G. Mitochondria of the protists. Annu Rev Genet 2004; 38:477-524.

35. Newmeyer DD, Ferguson-Miller S. Mitochondria: releasing power for life and unleashing the machineries of death. Cell 2003; 112:481-490.

36. Boxma B, de Graaf RM, van der Staay GW et al. An anaerobic mitochondrion that produces hydrogen. Nature 2005; 434:74-79.

37. Nasirudeen AM, Tan KS. Isolation and characterization of the mitochondrion-like organelle from Blastocystis hominis. J Micro Methods 2004; 58:101-109.

38. Muller M. The hydrogenosome. J Gen Microbiol 1993; 139:2879-2889.

39. Embley TM, van der Giezen M, Horner DS et al. Hydrogenosomes, mitochondria and early eukaryotic evolution. IUBMB Life 2003; 55:387-95.

40. Dyall SD, Johnson PJ. Origins of hydrogenosomes and mitochondria: Evolution and organelle biogenesis. Curr Opin Microbiol 2000; 3:404-411.

41. Dyall SD, Koehler CM, Delgadillo-Correa MG et al. Presence of a member of the mitochondrial carrier family in hydrogenosomes: Conservation of membrane-targeting pathways between hydrogenosomes and mitochondria. Mol Cell Biol 2001; 20:2488-2497.

42. Mai Z, Ghosh S, Frisardi M et al. Hsp60 Is Targeted to a Cryptic Mitochondrion-Derived Organelle ("Crypton") in the Microaerophilic Protozoan Parasite Entamoeba histolytica. Mol Cell Biol 1999; 19:2198-2205.

43. Ghosh S, Field J, Rogers R et al. The Entamoeba histolytica mitochondrion-derived organelle (crypton) contains double-stranded DNA and appears to be bound by a double membrane. Infect Immun 2000; 68:4319-4322.

44. Williams BAP, Hirt RP, Lucocq J et al. A mitochondrial remnant in the microsporidian Trachipleistophora hominis. Nature 2002; 418:865-869.

45. Putignani L, Tait A, Smith HV et al. Characterization of a mitochondrion-like organelle in Cryptosporidium parvum. Parasitology 2004; 129:1-18.

46. Katinka MD, Dupra S, Cornillot E et al. Genome sequence and gene compaction of the eukaryote parasite Encephalitozoon cuniculi. Nature 2001; 414:450-453.
47. Vivares C, Gouy M, Thomarat F et al. Functional and evolutionary analysis of a eukaryotic parasitic genome. Curr Opin Microbiol 2002; 5:499-505.
48. Lloyd D, Harris JC. Giardia: Highly evolved parasite or early branching eukaryote? Trends Microbiol 2002; 10:122-127.
49. Melino G, Knight RA, Nicotera P. How many ways to die? How many different models of cell death? Cell Death Diff 2005; 12:1457-1462.
50. Ameisen JC. On the origin, evolution, and nature of programmed cell death: a timeline of four billion years. Cell Death Diff 2002; 9:367-393.
51. André N. Hippocrates of Cos and apoptosis. The Lancet 2003; 361:1306.
52. Kerr JF, Wyllie AH, Currie AR. Apoptosis: a basic biological phenomenon with wide-ranging implications in tissue kinetics. Br J Cancer 1972; 26:239-257.
53. Yan N, Shi Y. Mechanisms of apoptosis through structural biology. Annu Rev Cell Dev Biol 2005; 21:35-56.
54. Bidère N, Su HC, Lenardo MJ. Genetic disorders of programmed cell death in the immune system. Annu Rev Immunol 2006; 24:321-52.
55. Martins LM. The serine protease Omi/HtrA2: a second mammalian protein with a Reaper-like function. Cell Death Diff 2002; 9:699-701.
56. Faccio L, Fusco C, Chen A et al. Characterization of a novel human serine protease that has extensive homology to bacterial heat shock endoprotease HtrA and is regulated by kidney ischemia. J Biol Chem 2000; 275:2581-2582.
57. Danial NN, Korsmeyer SJ. Cell Death: Critical Control Points. Cell 2004; 116:205-219.
58. Zamzami N, Larochette N, Kroemer G. Mitochondrial permeability transition in apoptosis and necrosis. Cell Death Diff 2005; 12:1478-1480.
59. Cande C, Cecconi F, Dessen P et al. Apoptosis-inducing factor (AIF): key to the conserved caspase- independent pathways of cell death? J Cell Sci 2002; 115:4727-4734.
60. Lamkan M, Declercq W, Kalai M et al. Alice in caspase land. A phylogenetic analysis of caspases from worm to man. Cell Death Diff 2002; 9: 358-361.
61. Green DR. Ten Minutes to Dead. Cell 2005; 121:671-674.
62. Chakravarti D, Hong R. SET-ting the stage for life and death. Cell 2003; 112:589-593.
63. Fan Z, Beresford PJ, Oh DY et al. Tumor suppressor NM23-H1 is a Granzyme A-activated DNase during CTL-mediated apoptosis, and the nucleosome assembly protein SET is its inhibitor. Cell 2003; 112:659-672.
64. Almgren MAE, Henriksson KCE, Fujimoto J et al. Nucleoside diphosphate kinase A/nm23-H1 promotes metastasis of NB69-derived human neuroblastoma CE. Mol Cancer Res 2004; 2:387-394.
65. Chowdhury D, Beresford PJ, Zhu P et al. The exonuclease TREX1 is in the SET complex and acts in concert with NM23-H1 to degrade DNA during granzyme A-mediated cell death. Mol Cell 2006; 23:133-42.
66. Goldtein P, Aubry L, Levraud J-P. Cell- death alternative model organisms: why and which? Nat Rev Mol Cell Biol 2003; 4:1-10.
67. Koonin EV, Aravind L. Origin and evolution of eukaryotic apoptosis: the bacterial connection. Cell Death Diff 2002; 9:394-404.
68. Yu YT, Snyder L. Translation elongation factor Tu cleaved by a phage exclusion system. Proc Natl Acad Sci USA 1994; 1:802-806.
69. Lewis K. Programmed Death in Bacteria. Microb Mol Biol Rev 2000; 64:503-514.
70. Berman-Frank I, Bidle K, Haramaty L et al. The demise of the marine cyanobacterium, Trichodesmium spp, via an autocatalyzed cell death pathway. Limnol Oceanogr 2004; 49:997-1005.
71. Ning SB, Guo HL, Wang L et al. Salt stress induces programmed cell death in prokaryotic organism Anabaena. J Appl Microbiol 2002; 93:15-28.
72. Bidle KB, Falkowski PG. Cell death planktonic photosynthetic microorganisms. Nat Rev Microb 2004; 2:643-655.
73. Arnoult D, Tatischeff I, Estaquier J et al. On the evolutionary conservation of the cell death pathway: mitochondrial release of an apoptosis-inducing factor during Dictyostelium discoideum cell death. Mol Biol Cell 2001; 12:3016-3030.
74. Kosta A, Roisin-Bouffay C, Luciani MF et al. Autophagy gene disruption reveals a nonvacuolar cell death pathway in Dictyostelium. J Biol Chem 2004; 12:48404-48409.
75. Kobayashi T, Endoh H. Caspase-like activity in programmed nuclear death during conjugation of Tetrahymena thermophila. Cell Death Diff 2003; 10:634-640.
76. Kobayashi T, Endoh H. A possible role of mitochondria in the apoptotic-like programmed nuclear death of Tetrahymena thermophila. FEBS J 2005; 272:5378-87.

77. Arnoult D, Akarid K, Grodet A et al. On the evolution of programmed cell death: apoptosis of the unicellular eukaryote Leishmania major involves cysteine proteinase activation and mitochondrion permeabilization. Cell Death Diff 2002; 9:65-81.
78. Alzate J, Alvarez-Barrientos A, Gonzàlez VM et al. Heat-induced programmed cell death in Leishmania infantum is reverted by Bcl-XL expression. Apoptosis 2006; 11:161-171.
79. Das M, Mukherjee SB, Shaha C. Hydrogen peroxide induces apoptosis-like death in Leishmania donovani promastigotes. J. Cell Sci 2001; 114:2461-2469.
80. Nguewa PA, Fuertes MA, Valladares B et al. Programmed cell death in trypanosomatids: a way to maximize their biological fitness? Trends Parasitol 2004; 20:375-379.
81. Piacenza L, Peluffo G, Radi R. L-arginine-dependent suppression of apoptosis in Trypanosoma cruzi: contribution of the nitric oxide and polyamine pathways. Proc Natl Acad Sci USA 2001; 98:7301-7306.
82. Welburn SC, Barcinski MA, Williams GT. Programmed cell death in trypanosomatids. Parasitol Today 1997; 13:22-6.
83. Fedorova ND, Badger JH, Robson GD et al. Comparative analysis of programmed cell death pathways in filamentous fungi. BMC Genomics 2005; 6:177.
84. Váchová L, Palková ZZ. Physiological regulation of yeast cell death in multicellular colonies is triggered by ammonia. J Cell Biol 2005; 5:711-717.
85. Herker E, Jungwirth H, Katharina et al. Chronological aging leads to apoptosis in yeast. J Cell Biol 2004; 164:501-507.
86. Al-Olayana EM, Williams GT, Hurd H. Apoptosis in the malaria protozoan, Plasmodium berghei: a possible mechanism for limiting intensity of infection in the mosquito. Int J Parasitol 2002; 32:1133-1143.
87. Deponte M, Becker K. Plasmodium falciparum—do killers commit suicide? Trends Parasitol 2004; 20: 165-169.
88. Brussaard CPD, Noordeloos AAM, Riegman R. Autolysis kinetics of the marine diatom Ditylum brightwellii (Bacillariophyceae) under nitrogen and phosphorus limitation and starvation. J Phycol 1997; 33:980-987.
89. Berges JA, Charlebois DO, Mauzerall DC et al. Differential effects of nitrogen limitation on photosynthetic efficiency of photosystems I and II in microalgae. Plant Physiol 1996; 110:689-696.
90. Vardi A, Berman-Frank I, Rozenberg T et al. Programmed cell death of the dinoflagellate Peridinium gatunense is mediated by $CO(2)$ limitation and oxidative stress. Curr Biol 1999; 9:1061-1064.
91. Franklin DJ, Berges JA. Mortality in cultures of the dinoflagellate Amphidinium carterae during culture senescence and darkness. Proc Biol Sci 2004; 271:2099-2107.
92. Berges JA, Charlebois DO, Mauzerall DC et al. Differential Effects of Nitrogen Limitation on Photosynthetic Efficiency of Photosystems I and II in Microalgae. Plant Physiol 1996; 110:689-696.
93. Segovia M, Haramaty L, Berges, JA. Cell death in the unicellular chlorophyte Dunaliella tertiolecta: a hypothesis on the evolution of apoptosis in higher plants and metazoans. Plant Physiol 2003; 132:99-105.
94. Chose O, Noel C, Gerbod D et al. A form of cell death with some features resembling apoptosis in the amitochondrial unicellular organism Trichomonas vaginalis. Exp Cell Res 2002; 276:32-39.
95. Chose O, Sarde CO, Noël C et al. Cell death in protists without mitochondria. Ann N Y Acad Sci 2003; 1010:121-125.
96. Tan KS, Nasirudeen AM. Protozoan programmed cell death—insights from Blastocystis deathstyles Trends Parasitol 2005; 21:547-550.
97. Rasoloson D, Vanacova S, Tomkova E. Mechanisms of in vitro development of resistance to metronidazole in Trichomonas vaginalis. Microbiology 2002; 148, 2467-2477.
98. Chose O, Sarde CO, Gerbod D et al. Programmed cell death in parasitic protozoans that lack mitochondria. Trends Parasitol 2003; 19:559-564.
99. Nasirudeen AM, Tan KS, Singh M et al. Programmed cell death in a human intestinal parasite, Blastocystis hominis. Parasitology 2001; 123:235-246.
100. Nasirudeen AM, Hian YE, Singh M et al. Metronidazole induces programmed cell death in the protozoan parasite Blastocystis hominis. Microbiology 2004; 150:33-43.
101. Adam RD. Biology of Giardia lamblia Clin Microbiol 2001; 14:447-475.
102. Benchimol M, Piva B, Campanati L. Visualization of the funis of Giardia lamblia by high-resolution field emission scanning electron microscopy—new insights. J Struct Biol 2004; 147:102-115.

CHAPTER 3

Programmed Cell Death and Trypanosomatids:
A Brief Review

Maria de Nazaré C. Soeiro* and Elen M. de Souza

Abstract

The phenomenon of apoptosis, one type of programmed cell death, is reviewed in three vector-borne trypanosomatids (*Trypanosoma cruzi*, *Trypanosoma brucei* and *Leishmania spp*) responsible for diseases of great medical and veterinary importance. Although some cytoplasmatic and nuclear apoptotic-like features of multicellular organisms such as phosphatidylserine exposure, cell retraction, nuclear condensation, DNA nicking, disruption of the mitochondrial membrane potential ($\Delta\Psi_m$) and caspase-like activity have been observed in these trypanosomatids, it still remains to be determined whether the type and pathways of apoptosis operating in these microorganisms are identical or not as in metazoans. Then, additional studies are essential to further characterize effector and regulatory molecules involved in trypanosomatid suicide program, which can provide the identification of new targets for future chemotherapeutic drug development and therapeutic interventions.

Programmed Cell Death

Cells die by different mechanisms that include active and passive processes such as, respectively, programmed cell death (PCD) or "cell suicide" and accidental necrosis. Necrosis is an unregulated process resulting from damage incurred to the cell, leading to plasma membrane and organelles disintegration and spreading of auto-antigens into the interstitium. PCD is a signal-dependent active process that requires specific gene transcription and protein synthesis leading to non inflammatory degradation of single cells that are aged, dysfunctional, unwanted or damaged by external stimuli.[1,2] In multicellular organisms, this genetically controlled cell suicide represents an integral part of the normal life since their growth, development and homeostasis are dependent on extensive cell renewal, while failures in this internal control may lead to cancerous growths and to the development of autoimmune diseases.[3,4] PCD represents an important aspect of host-pathogen interactions since microorganisms, such as bacteria, viruses and protozoan can manage the death programs of their host cells, activating, delaying or even preventing it.[5-8]

Although the onset and time course may differ with the cell type, once triggered, PCD is quite a similar process among different tissues types and species.[3] Based on some criteria such as morphology and environmental conditions different types of PCD have been characterized including apoptosis (type I-PCD), autophagy (type II-PCD), necrosis-like (type III-PCD) and

*Corresponding Author: Maria de Nazaré C. Soerio—Laboratorio de Biologia Celular, Dept. deUltra-estrutura e Biologia Celular, Instituto Oswaldo Cruz, FIOCRUZ, Av. Brasil 4365, 21040-900, Rio de Janeiro, RJ, Brazil. Email: soeiro@ioc.fiocruz.br

Programmed Cell Death in Protozoa, edited by José Manuel Pérez Martín.
©2008 Landes Bioscience and Springer Science+Business Media.

parapoptosis.[1,9-11] Although this classification provides a useful tool for the study of cell death, it does not fully satisfy the complex interplay between the different forms of PCD that may represent coexisting routes leading to the same endpoint.[1] Despite their individual particularities, these various forms of PCD share some biochemical as well as morphological features such as cell shrinkage, chromatin condensation and fragmentation, among others.

The Greek term "apoptosis" used for the first time by Kerr and collaborators[12] means "dropping off" petals and leaves from a healthy plant. This active cell death apparently does not involve own cells lysosomes and can be elicited by different triggers (serum or growth factor deprivation, chemical treatment, DNA damage, ultraviolet radiation, ionizing radiation, heat shock and oxidative stress, among others) as well as by different extracellular stimuli acting through cell-surface receptors.[11,13] Apoptosis is associated with distinctive biochemical and morphological changes, which include aberrant exposure of phosphatidylserine (PS) residues at the outer plasma membrane, chromatin condensation and marginalization in the nucleus, DNA cleavage into nucleosome-sized fragments, dissipation of the mitochondrial membrane potential ($\Delta\Psi_m$), cellular shrinkage, karyorrhexis and packing of cellular constituents into apoptotic vesicles, quickly removed by macrophages and/or neighboring cells.[14-19] In mammals, the removal of apoptotic bodies occurs through redundant receptors and the recognition of PS residues can lead to the production of transforming growth factor β_1 (TGF-β_1) by the phagocytic cells,[16] which can down regulate the inflammatory response. Two striking differences that can be highlighted between accidental necrosis and apoptosis are that in the PCD (i) the tissue architecture is mostly preserved and (ii) the DNA and other cell contents of the dying cells do not spill into the extracellular space and thus does not elicit inflammatory response.[4,12,20]

Apoptosis can be provoked (proapoptotic regulators) or inhibited (antiapoptotic regulators) by several signals or mediators, including Fas ligand, perforin, tumor necrosis factor-alpha, nitric oxide (NO) and intracellular proteases such as those from the caspase family.[21,22] The mitochondrion seems to play a central role in the cellular demise being considered a key regulator of the decision between life and death, since it not only receives and coordinates death signals but can also generate them.[1] In this context, the Bcl-2 family regulates the mitochondrion permeability and morphology acting as agonists (e.g., Bax or Bak) or as antagonists (e.g., Bcl-2) of the cell death. When Bax predominates the result is PCD, when Bcl-2 predominates the cell survives.[21]

Moreover, mitochondrion is an important source of reactive oxygen species (ROS), important inducers of apoptosis.[23] NO is another important regulatory and microbicidal molecule playing a role in the PCD, acting as a pro-apoptotic or an anti-apoptotic regulator depending on which PCD mechanism has been activated.[18,24] Another intrinsic signaling pathway involves the endoplasmic reticulum (ER) through the cleavage and activation of caspase-12 by calpain.[25] However, recent data demonstrated that ER Ca^{2+} depletion can trigger apoptosis independently of caspase-12.[26]

Programmed Cell Death in Unicellular Eukaryotes

In the past, based on the specific association of PCD and multicellularity it was hard to envisage that unicellular organisms could initiate a cell suicide program. However accumulated evidence shows that under a variety of stimuli, single-celled organisms such as yeasts,[27] dinoflagellates,[28] bacteria[29,30] as well as protozoa[9,31,32] can undergo a process analogous to PCD with apoptotic-like features, suggesting that this might be an ancient mechanism previous to the advent of multicellularity.[17,33] However, despite some phenotypic similarities, recent approaches showed differences regarding the induction and execution phases between apoptosis-like death of unicellular organisms and metazoans.[34,35]

Although the occurrence of PCD in single-celled organisms seems to be counterintuitive, several unicellulars do not live isolate but instead interact with each other as complex societies: as suggested,[36] PCD could occur in any living cells that display features of an organized network, which operates through interactions within themselves and/or with elements of their environment. It has been proposed that a suicide program could be developed in these organisms to control their growth and promote the clonality within their population.[37,38]

Programmed Cell Death in Trypanosomatids

Trypanosomatids, a large group of flagellated protozoa that are among the earliest branching eukaryotes, are distributed in different genera, including monogenetic and digenetic parasites of plants, insects and vertebrates. Morphological and biochemical features of apoptosis or type I-PCD have been reported in these organisms upon treatment with various stimuli, such as, hydrogen peroxide (H_2O_2), staurosporine (ST), heat shock, NO exposure, nutritional deprivation, chemotherapic agents, as well as during their intracellular development in their hosts.[6,34,37-41] The most common apoptotic features noted in these organisms include decreased $\Delta\Psi_m$, cellular retraction, caspase-like activity, release of cytochrome c, DNA fragmentation and PS exposure.

It has been proposed that PCD characteristics in a subpopulation of these parasites could represent an efficient adaptive strategy during the establishment of the host-parasite relationship.[34,37] Alternatively, it has been suggested that PCD in trypanosomatids could be part of a remnant process derived from ancestral death machinery over the course of the eukaryotic evolution, lacking a distinct function until triggered in response to diverse stimuli and stress conditions.[34] PCD in these unicellular microorganisms could also represent (i) equilibrium of host-parasite relationship by regulating the size of parasite populations, when there are insufficient nutrients or to avoid host death caused by a large number of parasites,[32,37] and (ii) evasion mechanism contributing to the down regulation of host immune response, avoiding an intense inflammatory reaction and/or contributing to the development of an aberrant-T-cell response.[5,39] Apoptotic-like death in these unicellular organisms could represent, at least in part of the population, an altruistic behavior where the dying of less competent or unviable individuals contribute for the success of the more competent ones that are able to be transmitted to the next hosts.[37]

Similarly to the role of mitochondria in multicellular organisms, it has been proposed that the unique mitochondrion-kinetoplast complex may play a major role type I-PCD in trypanosomatids.[34] The viability of the single mitochondrion in trypanosomatids is a vital biological aspect as compared with other organisms that display numerous mitochondria, since the presence of several copies of this organelle ensures some compensation for the injured ones. Then, the survival of organisms that display one mitochondrion would be directly related to the proper functioning of this single organelle.[42]

Two genera of the trypanosomatid family, *Leishmania* and *Trypanosoma*, display complexes life cycles with multiple differentiation forms alternating between invertebrate and vertebrate hosts.[43,44] The African trypanosome, *Trypanosoma brucei*, the American trypanosome, *Trypanosoma cruzi* and various species of *Leishmania* are transmitted by insect vectors and cause a variety of important diseases in humans and other mammals, being responsible for considerable human mortality and morbidity.[45] Thus, in view of their medical relevance and the fact that much of present knowledge regarding the mechanisms of programmed cell death that operate in trypanosomatids comes from studies performed with these parasites, the present review will deal with the current data regarding type I-PCD, the best studied form of programmed cell death in *Leishmania*, *T. brucei* and *T. cruzi*.

Leishmania spp

These parasites cause a wide range of human diseases that have nearly a worldwide distribution and results in high morbidity and mortality affecting 12 million people, with 30 million at risk of contracting the disease.[46] Depending on the parasite species and host susceptibility, human infections may manifest in localized self-healing cutaneous lesions to fatal visceral infections.[47] The parasite is transmitted to the vertebrate hosts by the bite of an infected female sand fly of the subgenera *Phlebotomus* (Old World) and *Lutzomia* (New World) that inoculates infective metacyclic promastigotes.[43] After being internalized by host macrophages, promastigotes differentiate into nonmotile amastigotes that are able to proliferate inside the phagolysosomes until the lyses of the infected cell, which releases large amounts of parasites that can be taken up by other host cells. The life cycle is completed when female phlebotomine takes the infecting blood meal containing amastigotes, which are converted to promastigotes in the insect midgut.[48,49] Inside the gut of the vector, the procyclic promastigotes multiply and some differentiate into several intermediate forms and ultimately into infectious metacyclic forms. However inside the sand fly gut, not all the promastigotes differentiate into metacyclic forms and it has been proposed that the remaining procyclic forms may undergo PCD in a similar way as it occurs with promastigotes at the stationary phase cultures in vitro, whose PCD is delayed by transferring these later forms to fresh media.[7,37] Some hypothesis regarding PCD in these stationary phase parasites have been suggested, including that PCD is triggered during parasite differentiation; by deprivation of nutrients essential for growth and/or by increased parasite density.[35] Moreover, it has been hypothesized that Leishmania PCD in vivo could be triggered in response to host mediators, representing an altruistic cell sorting mechanism to guarantee that some selected non-apoptotic amastigotes will be able to colonize in the next host keeping a silent chronic infection by silencing the host immune response.[37] Furthermore, it has been proposed that during infection of vertebrate hosts, the numerous intracellular forms localized within the parasitophorous vacuole may behave as a pseudo-multicellular organism, which under exposure to host cytotoxic agents would lead to the suicide of some sensitive parasites, ensuring the survival of those more resistant that could perform the transmission to the next host.[34] In fact, host-defense products like ROS, H_2O_2 and NO induce apoptosis in metazoans[50] as well as in promastigotes of *L. donovani*, which displayed oligonucleosomal DNA fragmentation, caspase-like activity, cleavage of a poly(ADP)ribose polymerase (PARP)-like protein and reduction of cell volume.[51]

Promastigotes of *L. (L.) amazonensis* when shifted from their optimal in vitro growth temperature (22-28°C) to that of the mammalian hosts (34-37°C), underwent a calcium-dependent apoptosis-like death, displaying DNA laddering and chromatin condensation.[52] Regarding the different evolutive forms, a comparative study showed that stationary axenic amastigotes of *L. donovani* displayed lower levels of I-PCD (monitored by nuclear condensation, DNA fragmentation and caspase-like activity) as compared to axenic promastigotes, suggesting that the later forms may be more susceptible to die by PCD in response to nutritional deprivation.[7]

Apoptotic-like parasites were detected not only in axenic conditions, but also interiorized in host cells. In vivo experiments with infected mouse[35] and hamster[53] indicated that this phenomenon is not an in vitro artifact but a natural process, which may regulate parasite growth during infection.

When *L. major* promastigotes were treated with staurosporine (ST), a protein kinase inhibitor known to induce apoptosis in mammalian cells, several cytoplasmic and nuclear features of apoptosis, including cell shrinkage, PS exposure with maintenance of plasma membrane integrity, disruption of $\Delta\Psi_m$ and cytochrome *c* release, nuclear chromatin condensation, DNA fragmentation and caspase-like activity were observed.[54] The cytoplasmic extract of ST-treated parasites led to DNA degradation of human isolated nuclei, which could be prevented by the cysteine protease inhibitor, E64, showing the involvement of parasite proteases and suggesting that, at least, part of the apoptotic machinery operating in *L. major* involves both cysteine proteinases and the mitochondrion, as described in mammalian cells.[54] On the other hand, although I-PCD features such as genomic DNA fragmentation into oligonucleosomes could be noted in *L. major*, *L. mexicana* and *L. amazonensis* maintained under serum deprivation, heat shock or NO exposure,[55] DNA fragmentation neither dependent on active caspases nor on the lysosomal cathepsin L-like enzymes was also reported in parasites not submitted to these stress conditions.[35]

Many drugs have been reported to induce apoptotic-like death in trypanosomatids. Miltefosine (MF), an alkylphosphocholine effective against human visceral leishmaniasis, triggers I-PCD-like in all forms of *L. donovani*.[56] Dying promastigotes treated with MF showed cytoplasmic, nuclear and membrane alterations similar to those presented by apoptotic metazoan cells, including cell shrinkage, DNA fragmentation into oligonucleosome-sized fragments and PS exposure.[57] Proteases seems to be part of the cell death machinery in MF-treated *L. donovani* since nuclear events, such as DNA digestion, were under the control of proteases sensitive to caspase inhibitors.[57]

Type I-PCD-like death (with DNA ladder formation) has also been reported after the treatment of *L. donovani* with the pentavalent antimony Pentostam (GlaxoSmithKline, UK) used for the treatment for human leishmaniasis.[7] High levels of caspase-like activity noted after the treatment of amastigotes with this drug, but not promastigotes, suggests differences in the PCD machinery or/and in the sensitivities to Pentostam between the two evolutive forms.[7] However, DNA fragmentation without caspase-like activity has been reported during the incubation of axenic amastigotes of *L. infantum* with the antimonial compound SbIII,[58] corroborating the hypothesis that both caspase-dependent and independent PCD pathways may occur in *Leishmania*. Amphotericin B, another anti-leishmanial drug, induced apoptosis-like death in late log phase promastigotes and axenic amastigotes of *L. donovani*, that displayed depolarization of mitochondrial membrane potential followed by induction of caspase-like activity.[7] Apoptotic-like parasites were also noted when *L. tropica*-infected macrophages were treated in vitro with both pentostam and amphotericin B[59] as well as in *L. donovani* lodged within macrophages isolated from treated patients.[60] Additionally, the treatment of macrophages infected with *L. donovani* with potassium antimony tartrate also led to intracellular parasites exposing PS and displaying DNA fragmentation.[61] The death was related with increased Ca^{2+} concentrations in both parasite and host cells and the antimony salt induced ROS production (primarily concentrated within the parasitophorous vacuoles) that was directly related to the leishmanial apoptotic-like features.[61] Trivalent antimony triggered extensive DNA fragmentation in both axenic and intracellular amastigotes of *L. (L.) amazonensis* in vitro.[55] Later, the same group reported that NO induced apoptosis-like death in amastigotes localized within canine macrophages exposed to autologous lymphocytes of dogs immunized with purified excreted-secreted antigens of *L. infantum*.[62]

DNA topoisomerases of *Leishmania* have been considered important targets for therapeutic treatment and some reports showed a correlation between the enzyme inhibition and apoptosis-like death. Camptothecin, an I DNA topoisomerase inhibitor, triggered mitochondrial dysfunction leading I-PCD-like death in both intracellular amastigotes and extracellular promastigotes of *L. donovani*.[23] The drug caused depolarization of the mitochondrial membrane followed by the activation of caspase-like proteases of the parasites after the release of cytochrome *c* into their cytosol.[23] A pentacyclic triterpenoid, dihydrobetulinic acid, also induced apoptosis-like death in both forms of the parasite by targeting DNA topoisomerases (I and II) and preventing enzyme-DNA binary complex formation.[63] The plant derived flavonoids luteolin and quercetin inhibited the growth of promastigotes of *L. donovani* arresting their cell cycle progression leading to apoptosis-like death by inducing topoisomerase II–mediated kinetoplast DNA cleavage.[64] The treatment of wild type and arsenite-resistant promastigotes of *L. donovani* with novobiocin, a catalytic inhibitor of topoisomerase II, caused morphological and biochemical features of metazoan apoptosis such as PS externalization, cytochrome C release to cytoplasm, activation of caspase-like activity, oligonucleosomal DNA fragmentation and in situ labeling of condensed and fragmented nuclei.[65] The analysis of the DNA from wild type and sodium arsenite-resistant strains of *L. donovani* submitted to paclitaxel and trifluralin treatment revealed an apoptosis-like death in response to these drugs.[66] Pentamidine (PT) is one of the best studied aromatic diamidine with a broad spectrum of anti-parasitic activity,[67] and recently it was reported that at least part of its antileishmanial activity could be attributed to targeting of parasite's toposoimerases, potentially at both nucleus and kinetoplast levels.[68] The concurrent inhibition of respiratory chain complex II during treatment of *L. donovani* promastigotes with PT increased the anti-parasitic activity of the drug and the generation of ROS and Ca^{2+} influx, which seemed to be critical events mediating apoptosis-like death.[42] Two other aromatic diamidines (DB75, commonly referred as furamidine and DB569, its phenyl substituted analogue) induced significant levels of apoptosis-like death in promastigotes of *Leishmania (L.) amazonensis* including cell retraction (Fig. 1) and alterations of $\Delta\Psi_m$ (Fig. 2), similarly as reported for *T. cruzi*.[41]

The exposure of PS in the surface of viable (propidium iodide-negative) infective amastigotes mediating their binding to specific host cell receptors have been implicated in the successful infectivity of *Leishmania (L.) amazonensis*.[69] The proposed model is that the recognition of PS on the surface of amastigotes by phagocytes could lead to TGF-β_1 release by the infected macrophages, which would stimulate the arginase activity of the infected and surrounding cells and consequently reduction of NO production and increase in polyamine synthesis by the host cells.[32]

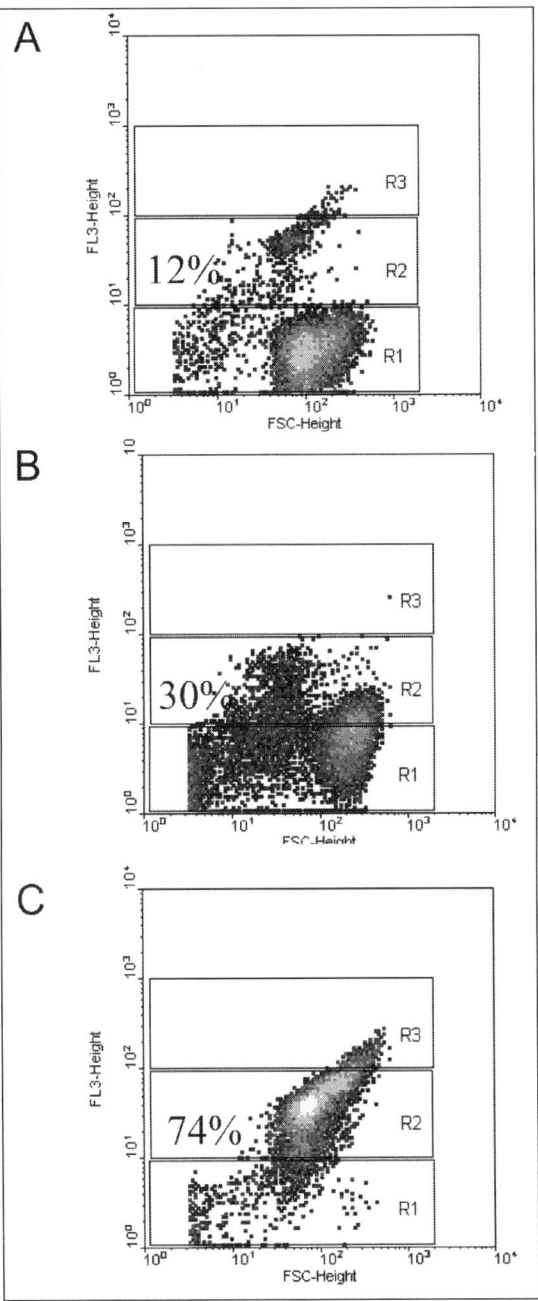

Figure 1. Flow cytometry analysis of PCD-I in *Leishmania (L.) amazonensis*: A) untreated pro-mastigotes; B) parasites treated with 4 μM DB-569 (24 h/37°C); C) apoptotic parasites (positive control) induced by high temperature (2 h/56°C). Untreated and treated parasites were collected, washed and then stained for 15 min with 100 μg/ml 7AAD that determines (R1) viable, (R2) apoptotic and (R3) necrotic parasites populations. See the enrichment of the R2 population after the aromatic diamidine treatment (30%) as compare to untreated parasites (12%).

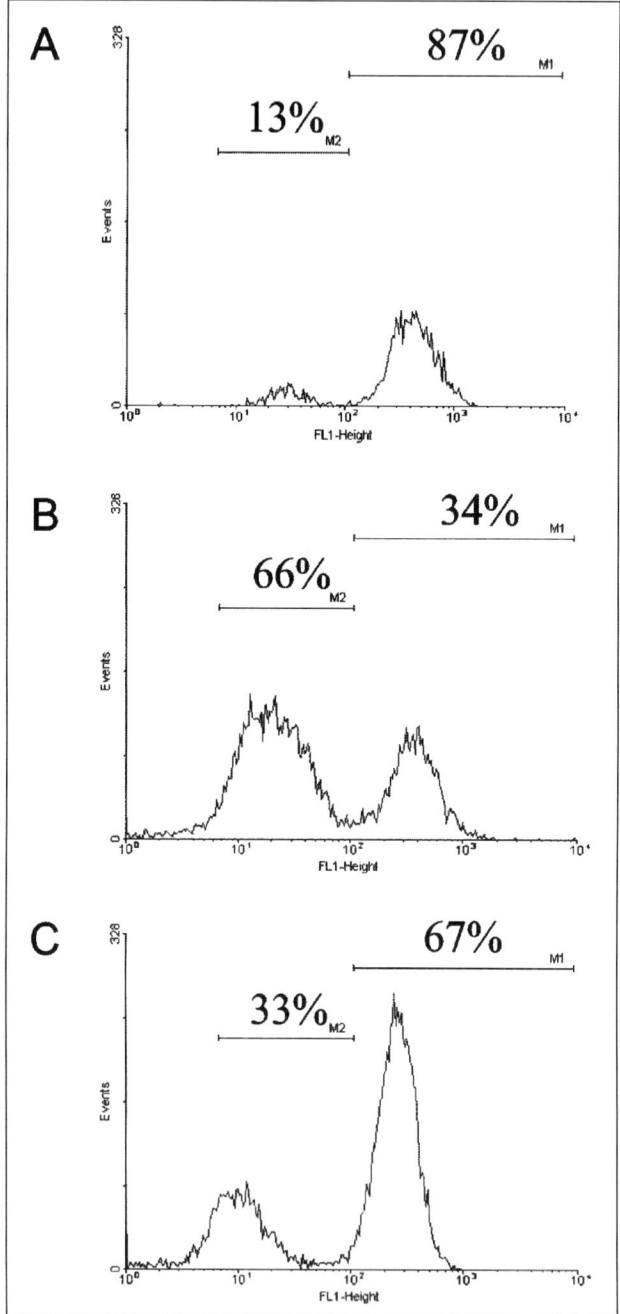

Figure 2. Analysis of the mitochondrial membrane potential ($\Delta\Psi_m$) of *Leishmania (L.) amazonensis* promastigotes by flow cytometry. Histograms show the fluorescence intensity of (A) untreated, (B) 4 μM DB569-treated and (C) staurosporine-treated parasites after their incubation with $DIOC_6$. The high fluorescence intensity peaks are marked as M1, whereas the low fluorescence intensity peaks are marked as M2 that represent the dissipation of $\Delta\Psi_m$.

In fact, the pretreatment of the parasites with annexin V, a PS ligand, inhibited their infectivity and reduced the TGF-β_1 production by infected macrophages.[69] Although the exact mechanism by which PS exposure increased amastigote infectivity is not fully understood, it may represent a natural phenotype displayed by a subpopulation of the parasites, enhancing the survival rates of the whole population.[70] Amastigotes of *L. (L.) amazonensis* obtained from BALB/c mice exposed higher levels of PS and were more infective as compared to those parasites derived from C57BL/6 mice, which expressed lower levels of PS and were less infective in vivo as well as in in vitro.[70] The parasites could be internalized in macrophages by macropinocytosis, whose activity was dependent on the amount of surface PS displayed by the parasites. The bulk of data suggest that PS exposure at the cell surface of viable amastigotes may contribute to the parasite invasion process besides inhibiting macrophage activity.[32,69,70]

As discussed above, the ability of *Leishmania* to undergo an apoptosis-like phenotype may contribute to (1) parasite evasion mechanisms from the immune response and (2) parasite growth in the infected phagocytic cells.[54] On the other hand, it has been suggested that the induction of apoptosis-like death in amastigotes may also represent a host mechanism to eliminate the parasite and/or control the disease, being an advantage factor to the host rather than to the parasite.[71] A recent finding that thioglycolate-elicited macrophages of BALB/c mice were unable to eliminate *L (L.) amazonensis* amastigotes, these phagocytes were able to kill intracellular forms from *L. guyanensis* inducing apoptosis-like death independent of NO but mediated by ROS, suggesting that this mechanism may account for the animals resistance to infection by the latter parasite.[71]

African Trypanosomes

Sleeping sickness is a severe disease, fatal if untreated and about 60 million people are at risk.[46] These protozoan parasites infect their hosts via the bite of blood feeding tsetse flies *(Glossina spp)* and cause a variety of diseases in humans and animals in Africa. While *T. brucei rhodesiense* is prevalent mainly in East and South Central Africa, *T. brucei gambiense* is found mostly in West and Central Africa. *T. b. brucei* is responsible for the cattle disease, but does not infect humans.[72] The human disease occurs in two forms: a chronic form caused by *T. b. gambiense* and an acute one caused by *T. b. rhodesiense*. Individuals suffering from either form of sleeping sickness die without treatment.[72,73]

Similar morphological features normally associated with I-PCD in metazoan cells were described in procyclic forms of *T. b. rhodesiense* that failed to establish mid gut colonization and died inside the vector.[74] Such death could be prevented by inhibiting their binding to lectins found at the tsetse mid gut, suggesting that parasites may be stimulated to follow an active death program by host factor(s).[75] Treatment of procyclic forms of *T. b. rhodesiense* with concanavalin A (Con A), wheat germ agglutinin and *Riccinus communis* agglutinin (specific for branched mannoses, GlcNAc β1,4 GlcNAc and βGal/βGalNAc, respectively) led to apoptotic-like morphological changes (associated with de novo gene expression) previously reported in *T. b. brucei* and *T. congolense*.[76,77] The incubation of procyclic forms with monoclonal antibodies (mAbs) specific for different epitopes of procyclin (their abundant surface protein) as well as the addition of the purified protein itself inhibited the Con A-induced cell death in *T. b. brucei*. In addition, mAbs specific for procyclin induced I-PCD in untreated parasites and mutant ones lacking all genes encoding procyclin became refractory to Con A and to anti-procyclin mAbs treatment.[78] The data suggested that procyclin molecules are involved in the Con A-induction of a novel form of cell death in trypanosomes, which has been defined as proto-apoptosis.[77] Besides typical type-I-PCD features, Con A-treated parasites also showed aberrant cell cycle events leading to multinucleation and altered cytokinesis that could represent a prelude to death.[78]

In *T. b. brucei*, ROS activate both Ca^{2+}-dependent and caspase-independent cell death pathways with some features of I-PCD found in mammalian cells. However the expression of mammalian anti-apoptotic regulator Bcl-2 in mutant parasites did not prevent apoptosis. In addition, nuclease activation in the apoptotic parasites was not due to caspase 3, caspase 1, calpain, serine protease, cysteine protease or proteasome activity.[79] These data suggest that although Ca^{2+} pathways may induce apoptosis-like death in the parasite, the specific proteins involved in this process might be distinct from those commonly found in metazoans.[79] In addition, the expression of the proapoptotic protein Bax in transgenic insect stages of *T. brucei* led to the release of cytochrome *c*, loss of oxidative phosphorylation, depolarization of the membrane potential and decrease of the intracellular ATP concentration. These events did not require other apoptotic stimuli and, contrasting to mammalian cells, were reversible.[80]

In procyclic forms of *T. b. rhodesiense*, I-PCD was correlated with the induction of several genes and the up-regulation of two of them, prohibitin (a proto-oncogene involved in cell cycle control, tumor suppression and senescence of higher eukaryotic cells) and TRACK (a member of the RACK family, intracellular receptor for activated protein kinase C), indicates similarities between cell suicide programs in multicellular eukaryotes and the parasites.[77,81] In addition, certain homologues of QM protein (a regulator of c-jun proto-oncoprotein) are differentially expressed when *T. brucei* parasites undergo PCD-like death,[82] corroborating the hypothesis regarding similarities between trypanosomatids and higher eukaryotes.[34]

In *T. brucei*, five genes encoding metacaspases (MTs) were identified. These cystein proteases display both caspase-like activity and present conserved caspase-like secondary structure.[83] The expression of *T. brucei* metacaspase 4 in yeast resulted in growth inhibition, mitochondrial alteration and clonal death.[83] However, although MTs show a limited similarity to the multicellular caspases, the present data do not provide any support regarding the direct involvement of *T. brucei* metacaspases in I-PCD.[84] Moreover, their colocalization with a small GTPase present in recycling endosomes suggests that they may have a function associated with RAB11 vesicles, which seems to be independent of the already known endocytic cell recycling process.[84] The authors concluded that since only little information is available regarding the function of MT, their possible role as metazoan caspases deserves further investigation.

Other apoptotic stimuli that have been reported to trigger the parasite apoptosis-like death include prostaglandin D2[85] and quercetin. This later is a potent immunomodulating flavonoid that induces the death of *T. b. gambiense*, without affecting normal human cell viability and its failure to induce I-PCD can represent a contributing factor to drug resistance in these trypanosomes.[86]

Trypanosoma cruzi

This parasite is the etiological agent of Chagas' disease, a zoonosis considered a major public health problem in the developing countries of Central and South America.[87] The parasite is transmitted among its hosts by hematophagous reduviid vectors and the overall prevalence of human infection is estimated at 16-18 million cases, with approximately 120 million people at risk of contracting the infection.[46] The myocardial damage is one of the most important pathological features responsible for the high morbidity and mortality rates in the infected people but the mechanisms that govern this pathogenesis are still poorly understood.[88] Primary cultures of heart muscle cells represent important tools for analyzing some aspects regarding the host-parasite cellular recognition as well as for other applications such as trypanocidal screenings in *T. cruzi* infected-cardiomyocytes (Fig. 3).

In *T. cruzi*, PCD seems to be involved in the control of parasite proliferation in axenic condition,[89] as well as when parasites are interiorized in mammalian host cells.[6] In the vertebrate hosts, the number of intracellular parasites could be enough to induce parasite type I-PCD, a phenomenon that may be associated to the parasite own growth control. Recent data showed that intracellular forms lodged within cardiac cells, fibroblast as well as peritoneal macrophages can display I-PCD features, including DNA fragmentation, both in vitro (Fig. 4) as well as in in vivo infections of mice[6] and dogs.[90] Interestingly, the rates of apoptotic-like death in intracellular forms of *T. cruzi* seems to vary according to the parasite biodemes: parasites from the Dm28c stock (representative of biodeme I, mostly involving the sylvatic cycle) displays less apoptotic-like levels as compared to the Y stock parasites (representative of biodeme II that predominates in the domestic cycle) that also presents lower proliferation rates in vitro.[6] In addition, type I-PCD in a sub-population of these parasites could represent a suitable strategy that permits their silent spreading, avoiding excessive damage to the host as well as preventing a host aggressive immune response against the parasites. In fact it has been shown that the uptake of apoptotic cells by phagocytic cells reduces the secretion of

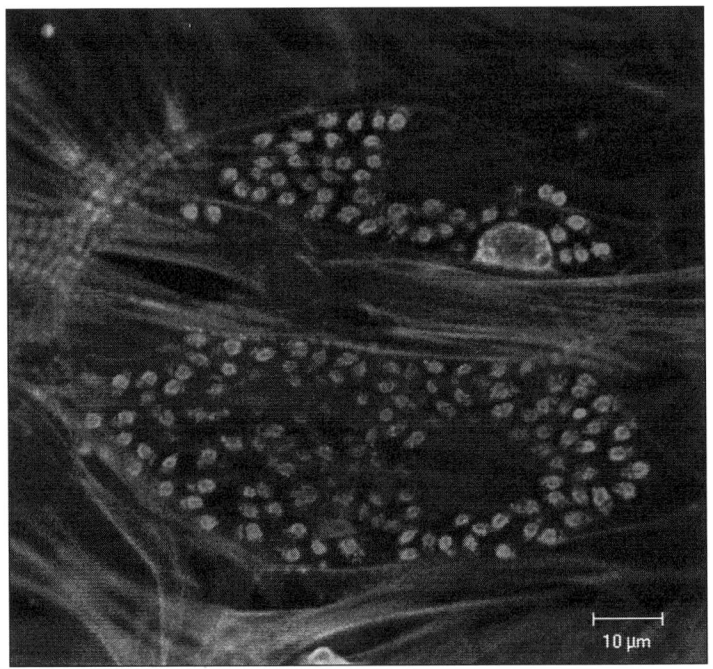

Figure 3. Confocal scanning laser microscopy showing primary cultures of embryonic cardiac cells after 72 h of infection with trypomastigotes of *Trypanosoma cruzi* (Y strain) in vitro. Double labeling of (i) cardiomyocytes with phalloidin-FITC, which labeled actin filaments (green) and (ii) intracellular parasites (red), which can be visualized by anti-*T. cruzi* antibodies revealing their intracellular distribution. A color version of this figure is available online at www.eurekah.com.

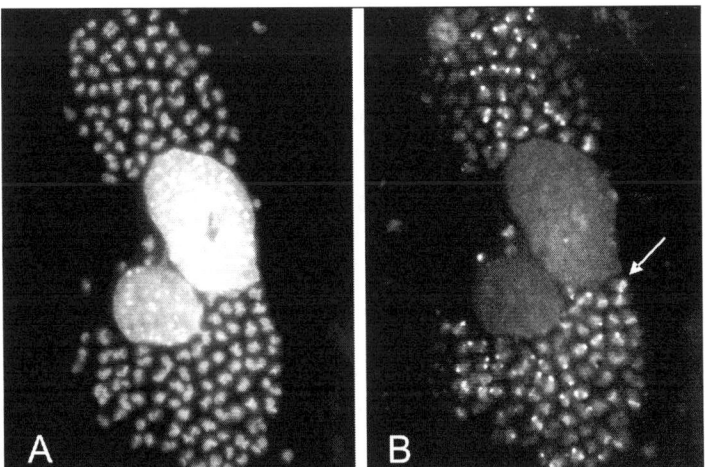

Figure 4. Fluorescent analysis by TUNEL technique of apoptotic cell death in intracellular forms of *T. cruzi* lodge within cardiomyocytes cultures (72 h of infection with Y strain). A) The host nuclei and both the nuclei and kinetoplast of the intracellular parasites labeled with 10 μg/ml DAPI; B) fragmented DNA in intracellular parasites noted by TUNEL staining.[6] Arrow indicates the intracellular parasites positive for TUNEL reaction.

pro-inflammatory cytokines and favors the intracellular growth of *T. cruzi*.[91] The I-PCD in these trypanosomatids could also represent a cell-sorting and altruistic mechanism for the selection of individuals that are fit to be transmitted to the next host.[37] Alternatively, it has been suggested that mammalian host cells could promote PCD, rather than necrosis of some trypanosomatids and the apoptotic-like features could be due to host cells products that deflagrate the cell suicide program in the intracellular parasites.[34] Epimastigotes, multiplicative forms found in the invertebrate host, can undergo apoptosis-like death triggered by different stimuli, including human serum as a source of complement.[92] The supplementation with *L*-arginine inhibited the DNA fragmentation and restored [^{3}H]-thymidine incorporation by means of (1) NO synthase-dependent NO production that suppresses apoptosis and (2) ADC-dependent production of polyamines that supports parasite proliferation.[92] The death of epimastigotes induced by human complement has been postulated as an evolutionary process where the non-adapted parasites could die to allow a host immunological silent state during the infection process.[89]

Interestingly apoptotic-like death of epimastigotes grown in axenic media can be either accelerated or prevented by modifying the culture conditions, suggesting that these parasites may use extracellular signals to regulate their survival/differentiation steps and subsequently control their cell density.[89] The few data regarding these aspects showed that in apoptotic ageing *T. cruzi* epimastigotes, the elongation factor 1-alpha (coded by housekeeping genes found in eukaryotes and prokaryotes) relocalize from the cytoplasm into the nucleus, suggesting that it can participate in the control of PCD in aged nonproliferating senescent parasites as has been reported for mammalian cells.[40,93]

PCD can be also triggered by the treatment of *T. cruzi* with *Bothrops jararaca* venom[94] and aromatic diamidines.[41] The snake venom inhibits the growth of epimastigotes causing mitochondrion swelling, kinetoplast disorganization and loss of its membrane potential, suggesting that the observed type I-PCD-like features may involve the mitochondrion cell death machinery.[94] The treatment of *T. cruzi* with the aromatic diamidines, DB75 and DB569 induced mitochondrial swelling, disruption of $\Delta\Psi_m$, PS exposure, kinetoplast disorganization and DNA fragmentation, which are characteristics of apoptosis-like death in multicellular eukaryotes.[41]

The genome of *T. cruzi* contains two genes with homology to those encoding MTs.[95] The proteins encoded by one of these genes (TcMCA3) are expressed in the four major developmental stages of the parasite, whereas the proteins encoded by gene TcMCA5 are only expressed in epimastigotes. Some indirect evidences suggest that MT might be involved in PCD-like of the parasite since the proteins coded by both genes change their subcellular localization during fresh human serum (FHS)-induced PCD, migrating into the nucleus and epimastigotes over-expressing TcMCA5 were more sensitive to FHS-induced PCD as compared to the controls. PCD was also paralleled by an increase in caspase substrate and the apoptotic nuclei cells were labeled in vivo with a pan-caspase fluorescent inhibitor.[95]

The occurrence of I-PCD in a sub-population of amastigotes localized within cardiac cells associated with the host cells' apoptosis could contribute for the parasite immune system evasion by promoting the synthesis of anti-inflammatory mediators deviating the aggressive immune response from the sites of the parasite proliferation and permitting its silent spreading.[6] The concomitant apoptosis of parasites and host cells could also contribute to the regulation of the cardiac lesions and parasite load, playing a role in chagasic autoimmunity.[5,6] A cell funeral model has been proposed for the pathogenic role of *T. cruzi* persistence in cardiac damage.[5] The authors suggested that apoptosis of cardiac cells and *T. cruzi* followed by the phagocytic clearance of apoptotic blebs (composed by parasite antigens and cardiac self-antigens) by immature dendritic cells concomitantly through scavenger receptors and phagocytic receptors could contribute to the establishment of an aberrant T-cell responses that may be responsible for persistence of *T. cruzi* and heart injury.[5]

It has been proposed that type I-PCD in trypanosomatids could act as a clone population controller where the altruistic death may favor the clone populations.[96] It has been proposed that the fluctuating parasitemia characteristic of protozoan infections in the mammalian bloodstream could also be resulted from the altruistic cell death program performed by these parasites in addition to the well known host immune response.[38] In fact, the nondividing trypomastigote forms of *T. cruzi* purified from the blood of infected mice exhibit apoptotic-like characteristics, possibly reflecting the ability of the bloodstream parasites to regulate their own numbers.[41]

Final Considerations

All the reported data suggest the existence of PCD in trypanosomatids, however the mechanisms, genes, pathways and molecules that are involved should be further explored, arguing for additional more extensive studies in order to better understand its role in these a single-celled organisms. This knowledge will contribute to identification of new targets, which can be studied for chemotherapeutic drug development and therapeutic interventions against parasitic infections, such as Chagas' disease, leishmaniasis and African trypanosomiasis. It has been proposed[34,37] that apoptotic features in trypanosomatids could (i) maximize their biological fitness in different cellular processes (such as parasite invasion, differentiation, proliferation control and escape from host immune response) and/or (ii) represent a process without a defined function, inherited through eukaryotic cell evolution, which might be triggered in response to diverse stress conditions and stimuli including pharmacological agents as we presently described. Thus, since up to now, the pathways, genes and proteins implicated in this active suicide program occurring in these unicellular organisms are still poorly understood, additional studies are imperative in order to identify effector and regulatory molecules of this kind of programmed cell death.

Acknowledgements

We thank Dr. Solange Lisboa de Castro and Dr. David Boykin for the careful revision and Dr. David Boykin for the aromatic diamidines. The present study was supported by grants from the Conselho Nacional Desenvolvimento Científico e Tecnológico (CNPq), DECIT/SCTIE/MS and MCT, by CNPq, PAPES IV/FIOCRUZ and FAPERJ.

References

1. Bras M, Queenan B, Susin SA. Programmed cell death via mitochondria: different modes of dying. Biochemistry 2005; 70:231-9.
2. Holdenrieder S, Stieber P. Apoptotic markers in cancer. Clin Biochem 2004; 37:605-17.
3. Häcker G. The morphology of apoptosis. Cell Tissue Res 2000; 301:5-17.
4. Mahoney JA, Rosen A. Apoptosis and autoimmunity. Curr Opin Immunol 2005; 17:583-8.
5. DosReis GA, Freire-de-Lima CG, Nunes MP et al. The importance of aberrant T-cell responses in Chagas disease. Trends Parasitol 2005; 21:237-43.
6. De Souza EM, Araujo-Jorge TC, Bailly C et al. Host and parasite apoptosis following Trypanosoma cruzi infection in in vitro and in vivo models. Cell Tissue Res 2003; 314:223-35.
7. Lee N, Bertholet S, Debrabant A et al. Programmed cell death in the unicellular protozoan parasite Leishmania. Cell Death Differ 2002; 9:53-64.
8. Lüder CG, Gross U, Lopes MF. Intracellular protozoan parasites and apoptosis: diverse strategies to modulate parasite-host interactions. Trends Parasitol 2001; 17:480-6.
9. Tan KSW, Nasirudeen AMA. Protozoan programmed cell death—insights from Blastocystis deathstyles. Trends Parasitol 2005; 21:547-50.
10. Sperandio S, de Belle I, Bredesen DE. An alternative, nonapoptotic form of programmed cell death. Proc Natl Acad Sci USA 2000; 97:14376-14381.
11. Clarke PG. Developmental cell death: morphological diversity and multiple mechanisms. Anat Embryol (Berl) 1990; 181:195-213.
12. Kerr JFR, Wyllie AH, Currie AR. Apoptosis: a basic biological phenomenon with wide-ranging implications in tissue kinetics. Br J Cancer 1972; 26:239-257.
13. Silva RD, Sotoca R, Johansson B et al. Hyperosmotic stress induces metacaspase- and mitochondria-dependent apoptosis in Saccharomyces cerevisiae. Mol Microbiol 2005; 58:824-34.
14. Tounekti O, Belehradek Jr J, Mir LM. Relationships between DNA fragmentation, chromatin condensation and changes in flow cytometry profiles detected during apoptosis. Exp Cell Res 1995; 217:506-516.
15. Daniel PT, Sturm I, Ritschel S et al. Detection of genomic DNA fragmentation during apoptosis (DNA ladder) and the simultaneous isolation of RNA from low cell numbers. Anal Biochem 1999; 266:110-115.
16. Fadok VA, Bratton DL, Rose DM et al. A receptor for phosphatidylserine-specific clearance of apoptotic cells. Nature 2000; 405: 85-90.

17. Ameisen JC. On the origin, evolution and nature of programmed cell death: a timeline of four billion years. Cell Death Differ 2002; 9:367-93.
18. Behnia M, Robertson KA, Martin WJ. Role of apoptosis in host defense and pathogenesis of disease. Ches 2000; 117:1771-1777.
19. Bossy-Wetzel E, Green DR. Apoptosis: checkpoint at the mitochondrial frontier. Mutat Res 1999; 434:243-251.
20. Lopes MF, DosReis GA. The macrophage haunted by cell ghosts: a pathogen grows. Immunol Today 2000; 21:489-494.
21. Chwieralski CE, Welte T, Buhling F. Cathepsin-regulated apoptosis. Apoptosis 2006; 11:143-9.
22. Cosulich SC, Savory PJ, Clarke PR. Bcl-2 regulates amplification of caspase activation by cytochrome c. Curr Biol 1999; 9:147-50.
23. Sen N, Das BB, Ganguly A et al. Camptothecin induced mitochondrial dysfunction leading to programmed cell death in unicellular hemoflagellate Leishmania donovani. Cell Death Differ 2004; 11:924-36.
24. Rachek LI, Grishko VI, Ledoux SP et al. Role of nitric oxide-induced mtDNA damage in mitochondrial dysfunction and apoptosis. Free Radic Biol Med 2006; 40:754-62.
25. Liu H, Baliga R. Endoplasmic reticulum stress-associated caspase 12 mediates cisplatin-induced LLC-PK1 cell apoptosis. J Am Soc Nephrol 2005; 16:1985-92.
26. Nakano T, Watanabe H, Ozeki M et al. Endoplasmic reticulum Ca^{2+} depletion induces endothelial cell apoptosis independently of caspase-12. Cardiovasc Res 2006; 69:908-15.
27. Gourlay CW, Ayscough KR. The actin cytoskeleton in ageing and apoptosis. FEMS Yeast Res 2005; 5:1193-8.
28. Vardi A, Berman-Frank I, Rozenberg T et al. Programmed cell death of the dinoflagellate Peridinium gatunense is mediated by CO_2 limitation and oxidative stress. Curr Biol 1999; 9:1061-4.
29. Lewis K. Programmed cell death in bacteria. Microbiol Mol Biol Rev 2000; 64:503-514.
30. Sat B, Hazan R, Fisher T et al. Programmed cell death in Escherichia coli: some antibiotics can trigger mazEF lethality. J Bacteriol 2001; 183:2041-5.
31. Tsuda A, Witola WH, Ohashi K et al. Expression of alternative oxidase inhibits programmed cell death-like phenomenon in bloodstream form of Trypanosoma brucei rhodesiense. Parasitol Int 2005; 54:243-51.
32. Wanderley JLM, Benjamin A, Real F et al. Apoptotic mimicry: an altruistic behavior in host/Leishmania interplay Braz J Med Biol Res 2005; 38:807-812.
33. Cornillon S, Foa C, Davoust J et al. Programmed cell death in Dictyostelium. J Cell Sci 1994; 107:2691-704.
34. Nguewa PA, Fuertes MA, Valladares B et al. Programmed cell death in trypanosomatids: a way to maximize their biological fitness? Trends Parasitol 2004; 20:375-80.
35. Zangger H, Mottram JC, Fasel N. Cell death in Leishmania induced by stress and differentiation: programmed cell death or necrosis? Cell Death Differ 2002; 9:1126-39.
36. DosReis GA, Barcinski MA. Apoptosis and parasitism: from the parasite to the host immune response. Adv Parasitol 2001; 49:133-61.
37. Debrabant A, Nakhasi. Programmed cell death in trypanosomatids: is it an altruistic mechanism for survival of the fittest? Kinetoplastid Biol Dis 2003; 2:7.
38. Welburn SC, Barcinski M, Williams G. Programmed cell death in trypanosomatids. Parasitol Today 1997; 13:22-26.
39. Barcinski MA, DosReis GA. Apoptosis in parasites and parasite-induced apoptosis in the host immune system: a new approach to parasitic diseases. Braz J Med Biol Res 1999; 32:395-401.
40. Ouaissi A. Apoptosis-like death in trypanosomatids: search for putative pathways and genes involved. Kinetoplastid Biol Dis 2003; 2:5.
41. De Souza EM, Menna-Barreto R, Araújo-Jorge TC et al. Antiparasitic Activity of Aromatic diamidines is related to apoptosis-like death in Trypanosoma cruzi. Parasitology 2006; 27:1-5.
42. Mehta A, Shaha C. Apoptotic death in Leishmania donovani promastigotes in response to respiratory chain inhibition: complex II inhibition results in increased pentamidine cytotoxicity. J Biol Chem 2004; 279:11798-11813.
43. Vannier-Santos MA, Martiny A, de Souza W. Cell biology of Leishmania spp. invading and evading. Curr Pharm 2002; 8:297-318.
44. De Souza W. From the cell biology to the development of new chemotherapeutic approaches against trypanosomatids: dreams and reality. Kinetoplastid Biol Dis 2002; 1:3.
45. WHO. The world health report. 1999; Genova. World Health Organization.
46. WHO. The world health report. 2002; Genova. World Health Organization.
47. TDR. http://www.who.int/tdr/diseases/leish/default.htm 2002.
48. Sacks D, Kamhawi S. Molecular aspects of parasite-vector and vector-host interactions in leishmaniasis. Annu Rev Microbiol 2001; 55:453-83.
49. Peters C, Aebischer T, Stierhof YD et al. The role of macrophages receptors in adhesion and uptake of Leishmania mexicana amastigotes. J Cell Scien 1995; 108:3715-3724.

50. Mignotte B, Vayssiere JL. Mitochondria and apoptosis. Eur J Biochem 1998; 252:1-15.
51. Das M, Mukherjee SB, Shaha C. Hydrogen peroxide induces apoptosis-like death in Leishmania donovani promastigotes. J Cell Science 2001; 114: 2461-2469.
52. Moreira ME, Del Portillo HA, Milder RV et al. Heat shock induction of apoptosis in promastigotes of the unicellular organism Leishmania (Leishmania) amazonensis. J Cell Physiol 1996; 167:305-13.
53. Lindoso JA, Cotrim PC, Goto H. Apoptosis of Leishmania (Leishmania) chagasi amastigotes in hamsters infected with visceral leishmaniasis. Int J Parasitol 2004; 34:1-4.
54. Arnoult D, Akarid K, Grodet A et al. On the evolution of programmed cell death: apoptosis of the unicellular eukaryote Leishmania major involves cysteine proteinase activation and mitochondrion permeabilization. Cell Death Differ 2002; 9:65-81.
55. Holzmuller P, Sereno D, Cavaleyra M et al. Nitric oxide-mediated proteasome-dependent oligonucleosomal DNA fragmentation in Leishmania amazonensis amastigotes. Infect Immun 2002; 70:3727-35.
56. Verma NK, Dey CS. Possible mechanism of miltefosine-mediated death of Leishmania donovani. Antimicrob Agents Chemother 2004; 48:3010-5.
57. Paris C, Loiseau PM, Bories C et al. Miltefosine induces apoptosis-like death in Leishmania donovani promastigotes. Antimicrob Agents Chemother 2004; 48:852-9.
58. Sereno D, Holzmuller P, Mangot I et al. Antimonial-mediated DNA fragmentation in Leishmania infantum amastigotes. Antimicrob Agents Chemother 2001; 45:2064-9.
59. Langreth SG, Berman JD, Riordan GP et al. Fine-structural alterations in Leishmania tropica within human macrophages exposed to antileishmanial drugs in vitro. J Protozool 1983; 30:555-61.
60. Chulay JD, Fawcett DW, Chunge CN. Electron microscopy of Leishmania donovani in splenic aspirates from patients with visceral leishmaniasis during treatment with sodium stibogluconate. Ann Trop Med Parasitol 1985; 79:417-429.
61. Sudhandiran G, Shaha C. Antimonial-induced increase in intracellular Ca^{2+} through nonselective cation channels in the host and the parasite is responsible for apoptosis of intracellular Leishmania donovani amastigotes. J Biol Chem 2003; 278:25120-25132.
62. Holzmuller P, Cavaleyra M, Moreaux J et al. Lymphocytes of dogs immunised with purified excreted-secreted antigens of Leishmania infantum co-incubated with Leishmania infected macrophages produce IFN gamma resulting in nitric oxide-mediated amastigote apoptosis. Vet Immunol Immunopathol 2005; 106:247-57.
63. Chowdhury AR, Mandal S, Goswami A et al. Dihydrobetulinic acid induces apoptosis in Leishmania donovani by targeting DNA topoisomerase I and II: implications in antileishmanial therapy. Mol Med 2003; 9:26-36.
64. Mittra B, Saha A, Chowdhury AR et al. Luteolin, an abundant dietary component is a potent anti-leishmanial agent that acts by inducing topoisomerase II-mediated kinetoplast DNA cleavage leading to apoptosis. Mol Med 2000; 6:527–541.
65. Singh G, Jayanarayan KG, Dey CS. Novobiocin induces apoptosis-like cell death in topoisomerase II over-expressing arsenite resistant Leishmania donovani. Mol Biochem Parasitol 2005; 141: 57-69.
66. Jayanarayan KG, Dey CS. Altered tubulin dynamics, localization and posttranslational modifications in sodium arsenite resistant Leishmania donovani in response to paclitaxel, trifluralin and a combination of both and induction of apoptosis-like cell death. Parasitology 2005; 131:215-30.
67. Soeiro MNC, De Souza EM, Stephens CE et al. Aromatic diamidines as antiparasitic agents. Exp Opinion Investig Drugs 2005; 14: 957-72.
68. Jean-Moreno V, Rojas R, Goyeneche D et al. Leishmania donovani: differential activities of classical topoisomerase inhibitors and antileishmanials against parasite and host cells at the level of DNA topoisomerase I and in cytotoxicity assays. Exp Parasitol 2006; 112:21-30.
69. De Freitas Balanco JM, Moreira ME, Bonomo A et al. Apoptotic mimicry by an obligate intracellular parasite downregulates macrophage microbicidal activity. Curr Biol 2001; 27:1870-3.
70. Wanderley JL, Moreira ME, Benjamin A et al. Mimicry of apoptotic cells by exposing phosphatidylserine participates in the establishment of amastigotes of Leishmania (L) amazonensis in mammalian hosts. J Immunol 2006; 176:1834-9.
71. Sousa-Franco J, Araujo-Mendes E, Silva-Jardim I et al. Infection-induced respiratory burst in BALB/c macrophages kills Leishmania guyanensis amastigotes through apoptosis: possible involvement in resistance to cutaneous leishmaniasis. Microbes Infect 2006; 8:390-400.
72. Denise H, Barrett MP. Uptake and mode of action of drugs used against sleeping sickness. Biochem Pharmacol 2001; 61:1-5.
73. Docampo R, Moreno SN. Current chemotherapy of human African trypanosomiasis. Parasitol Res 2003; 90:10-3.
74. Welburn SC, Maudlin I, Ellis DS. Rate of trypanosome killing by lectins in midguts of different species and strains of Glossina. Med Vet Entomol 1989; 3:77-82.
75. Maudlin I, Welburn SC. Maturation of trypanosome infections in tsetse. Exp Parasitol 1994; 79:202-5.
76. Welburn SC, Dale C, Ellis D et al. Apoptosis in procyclic Trypanosoma brucei rhodesiense in vitro. Cell Death Differ 1996; 3: 229-236.

77. Welburn SC, Lillico S, Murphy NB. Programmed cell death in procyclic form Trypanosoma brucei rhodesiense ¯identification of differentially expressed genes during conA induced death. Mem Inst Oswaldo Cruz 1999; 94:229-234.
78. Pearson TW, Beecroft RP, Welburn SC et al. The major cell surface glycoprotein procyclin is a receptor for induction of a novel form of cell death in African trypanosomes in vitro. Mol Biochem Parasitol 2000; 111:333-349.
79. Ridgley EL, Xiong ZH, Ruben L. Reactive oxygen species activate a Ca2+-dependent cell death pathway in the unicellular organism Trypanosoma brucei brucei. Biochem J 1999; 340:33-40.
80. Esseiva AC, Chanez A, Bochud-Allemann N et al. Temporal dissection of Bax-induced events leading to fission of the single mitochondrion in Trypanosoma brucei. EMBO Rep 2004; 5:268-73.
81. Welburn SC, Murphy NB. Prohibiting and RACK homologues are up-regulated in trypanosomes induced to undergo apoptosis and in naturally occurring terminally differentiated forms. Cell Death Differ 1998; 5:615-22.
82. Lillico SG, Mottram JC, Murphy NB et al. Characterisation of the QM gene of Trypanosoma brucei. FEMS Microbiol Lett 2002; 211:123-8.
83. Szallies A, Kubata BK, Duszenko M. A metacaspase of Trypanosoma brucei causes loss of respiration competence and clonal death in the yeast Saccharomyces cerevisiae. FEBS Lett 2002; 517:144-50.
84. Helms MJ, Ambit A, Appleton P et al. Bloodstream form Trypanosoma brucei depend upon multiple metacaspases associated with RAB11-positive endosomes. J Cell Sci 2006; in press Epub ahead of print.
85. Figarella K, Rawer M, Uzcategui NL et al. Prostaglandin D2 induces programmed cell death in Trypanosoma brucei bloodstream form. Cell Death Differ 2005; 12:335-46.
86. Mamani-Matsuda M, Rambert J, Malvy D et al. Quercetin induces apoptosis of Trypanosoma brucei gambiense and decreases the proinflammatory response of human macrophages. Antimicrob Agents Chemother 2004; 48:924-9.
87. Hirst SI, Stapley LA. Parasitology: the dawn of a new millennium. Parasitol Today 2000; 16:1-3.
88. Tarleton RL. Chagas disease: a role for autoimmunity? Trends Parasitol 2003; 19:447-51.
89. Ameisen JC, Idziorek T, Billaut-Multo O et al. Apoptosis in a unicellular eukaryote (Trypanosoma cruzi)—implications for the evolutionary origin and role of programmed cell death in the control of cell proliferation, differentiation and survival. Cell Death Differ 1995; 2:285-300.
90. Zhang J, Andrade ZA, Yu ZX et al. Apoptosis in a canine model of acute chagasic myocarditis. J Mol Cell Cardiol 1999; 31:581-596.
91. Freire-de-Lima CG, Nascimento DO, Soares MBP et al. Uptake of apoptotic cells drives the growth of a pathogenic trypanosome in macrophages. Nature 2000; 403:199-203.
92. Piacenza L, Peluffo G, Radi R. L-Arginine-dependent suppression of apoptosis in Trypanosoma cruzi: contribution of the nitric oxide and polyamine pathways. Proc Natl Acad Sci 2001; 98:7301-7306.
93. Billaut-Mulot O, Fernandez-Gomez R, Loyens M et al. Trypanosoma cruzi elongation factor 1-alpha: nuclear localization in parasites undergoing apoptosis. Gene 1996; 174:19-26.
94. Deolindo P, Teixeira-Ferreira AS, Melo EJ et al. Programmed cell death in Trypanosoma cruzi induced by Bothrops jararaca venom. Mem Inst Oswaldo Cruz 2005; 100:33-8.
95. Kosec G, Alvarez VE, Aguero F. et al. Metacaspases of Trypanosoma cruzi: Possible candidates for programmed cell death mediators. Mol Biochem Parasitol 2006; 145:18-28.
96. Al-Olayan EM, Williams GT, Hurd H. Apoptosis in the malaria protozoan, Plasmodium berghei: a possible mechanism for limiting intensity of infection in the mosquito. J Parasitol 2002; 32:1133-1143.

Programmed Cell Death in African Trypanosomes

Katherine Figarella, Néstor L. Uzcátegui, Viola Denninger, Susan Welburn and Michael Duszenko*

Abstract

Since the discovery of programmed cell death in multicellular organisms and due to its definition as a mechanism to maintain the individual haemostasis of cellular and organ integrity, it was not plausible to think that such a phenomenon could also occur in unicellular organisms. However, during the last decade considerable experimental evidence has accumulated that confirm the existence of programmed (i.e., genetically encoded) mechanisms of cell death in a wide variety of single-cell organisms, including bacteria as well as free living and parasitic protozoa. Moreover, the discovery of biofilm formation and quorum sensing in bacteria[1] and similar observations especially in protozoan parasites[2] has changed our perception not to view unicellular organisms as selfish, self-contained and autonomous entities but as well organized cell populations expressing established communication patterns, thus resembling their multicellular counterparts. In this chapter we will summarize the most obvious findings regarding programmed cell death in African trypanosomes.

General Features of *Trypanosoma brucei*

African trypanosomes are protozoan parasites that cause sleeping sickness in humans and Nagana in animals, especially in livestock. They are transmitted by the tsetse fly and thus restricted to the area of fly occurrence, i.e., the tsetse belt in central Africa. Human African trypanosomiasis (HAT) is a fatal disease, threatening about 60 million people in the endemic area with a prevalence of up to several hundred thousand new cases each year. Existing medical treatment relies on highly toxic arsenical (melarsoprol) or insufficient and yet expensive (eflornithine) drugs and compliance is therefore limited, especially as HAT effects mainly some of the poorest countries in the world. In some endemic foci HAT is currently the main killer which exceeds even HIV. Thus new, effective, cheap and easy to apply drugs are urgently needed.

Trypanosomes undergo a complex life cycle with formation of several genetically controlled populations to cope with environmental conditions in both hosts, mammals and the tsetse fly. From life in the mammalian body fluids (blood, lymph and liquor) at maintained 37°C with the luxury of a well controlled glucose supply, parasites enter the environment of the insect gut and need to adapt to cooler and inconstant temperatures, new food sources (mainly proline) and hostile digestive enzymes. During the adaptation process, trypanosomes undergo several morphological and metabolical changes leading to the development of distinctive populations (Fig. 1). The proliferative population in mammals shows a slender morphology, depends exclusively on glycolysis for energy production[3] and

*Corresponding Author: Michael Duszenko—Department of Biochemistry, University of Tuebingen, Hoppe-Seyler-Str. 4, 72076 Tuebingen, Germany.
Email: michael.duszenko@uni-tuebingen.de

Programmed Cell Death in Protozoa, edited by José Manuel Pérez Martín.
©2008 Landes Bioscience and Springer Science+Business Media.

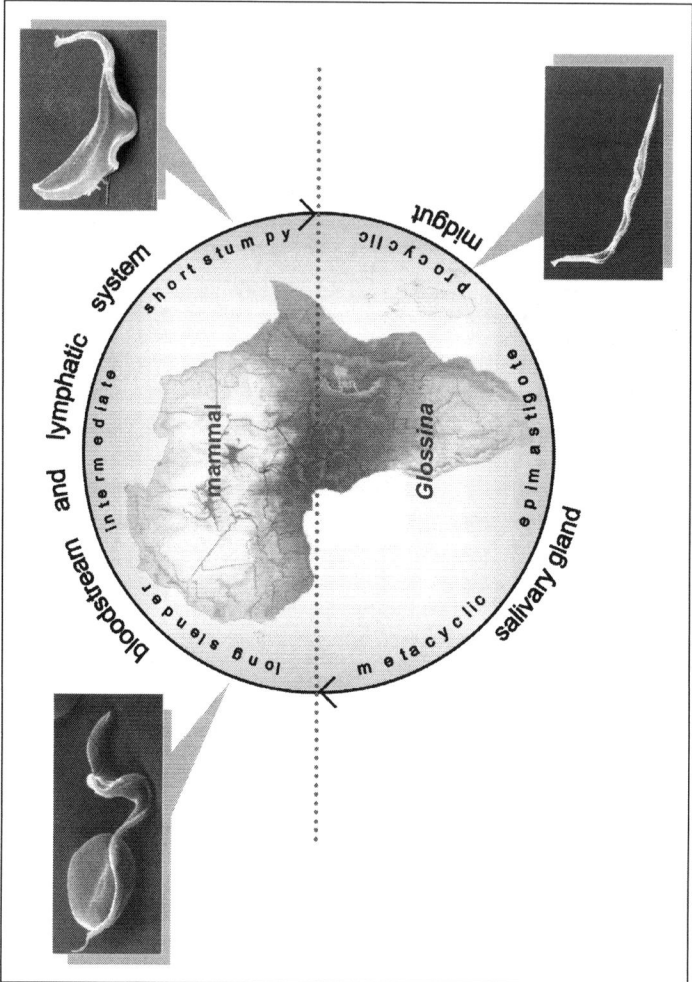

Figure 1. Life cycle of *Trypanosoma brucei*. Scanning electron microscopy pictures show the characteristic morphology of long slender, short stumpy and procyclic forms.

undergoes antigenic variation to escape the host's immune system.[4] Dependent on cell density and due to quorum sensing, part of the slender population undergoes a terminal differentiation to a non dividing population of stumpy morphology.[5] This form contains a higher but still not fully developed mitochondrion and is thus pre-adapted to live within the insect vector; it does no longer undergo antigenic variation and dies if it is not taken up by the tsetse fly.[6] In the mammalian host, the surface of *T. brucei* is covered by a dense coat consisting of about 10^7 copies of a single protein, the variant surface glycoprotein (VSG). By sequential expression of genetically encoded variants of VSG molecules, trypanosomes evade the immune response, because at the time antibodies have been developed, parasites with an antigenically different VSG coat have appeared. This mechanism of antigenic variation leads to characteristic parasitaemic waves as parasites expressing the "old" coat are cleard from blood while the new variant grows to higher density.

After transmission to the tsetse fly upon a blood meal, slender parasites will die after glucose is consumed, while stumpy parasites transform into the procyclic form which contains a fully functional mitochondrion. Following an intricate track in the fly, parasites reach the salivary glands where an epimastigote population develops which finally undergoes terminal differentiation to the mammal-infective metacyclic stage. This complex life cycle needs to be regulated to guarantee parasite's transmission but also to prevent uncontrolled growth in either host. Until some years ago it seemed logical to assume that the parasitaemia within the mammalian host was controlled only by the appearance of specific antibodies again VSG.[7] However, parasitic waves have also been detected in vitro if conditioned culture medium was regularly replaced by fresh medium.[8] In addition, observations in different laboratories indicate that differentiation in trypanosomes depends on quorum sensing in response to a differentiation inducing factor of a yet unknown chemical structure.[8,9] Quorum sensing is not restricted to trypanosomes but widespread among unicellular populations and functionally related factors have been isolated and analyzed e.g., from bacteria (homoserine lactone[10]), dictyostelium (a chlorinated aromatic compound[11]) and Plasmodia (xanthurenic acid[12]).

Programmed Cell Death in Multicellular and Unicellular Organisms

Programmed cell death (PCD) is a fundamental biological process that is implicated in early development such as during metamorphosis in insects and amphibians and organogenesis in virtually all multicellular organisms. It has been described in metazoan organisms as a form of self-regulation and a system of social control on cell populations.[13] Disorders in the execution of this process such as an elevated cell death can contribute to neurodegenerative diseases like Alzheimer and Parkinson syndromes, ischaemic injury, etc. On the other hand, insufficient cell death can lead to cancer, persistent viral infection, or autoimmune disorders. So far several forms of programmed cell death in multicellular organisms have been described: apoptosis, caspase-independent apoptosis also known as apoptosis-like, paraptosis, autophagy, necrosis-like, etc (Fig. 2a). The molecular mechanisms that trigger these different PCD forms are specific for each one. The best described form of PCD is classical apoptosis, which is characterized by an intrinsic or extrinsic activation of procaspases by limited proteolysis (Fig. 2b). The term apoptosis has been coined to define specifically the removal of unwanted cells for the benefit of the whole organism (like in embryogenesis or maturation of the immune system) and can be seen as an altruistic suicide of singular cells which otherwise may be harmful for the whole organism, and degrade without induction of inflammatory reactions.

In the case of unicellular organisms, experimental evidence has accumulated over years which indicate that many of these organisms may undergo forms of cell deaths showing typical hallmarks of apoptosis, like cell shrinkage, chromatin condensation, DNA degradation, membrane disturbances as evidenced by phosphatidylserine exposition in the outer leaflet of the plasma membrane or loss of the mitochondrial membrane potential (ψ_m) and others. Extensive discussions centered around the question why unicellular organisms may have developed a molecular machinery to induce cell death and if they are able to exhibit an altruistic behavior. However, in the course of investigations it is turning obvious that some form of programmed cell death must occur here as well. Indeed, observations such as growth arrest either due to nutritional deprivation or by drugs, inhibition of signaling molecules, use of oxidative stress or heat shock, aging, differentiation and mutations in cell cycle regulating genes have been reported to induce PCD phenotypes in cells ranging from bacteria to unicellular eukaryotes.[14] Interestingly, it has been suggested that functions of PCD in unicellular organisms can be similar to those found in metazoa, since these organisms may interact with themselves as organized communities.[15-17] However, the cross-talk among cells and the signal cascade that is generated during these cell deaths processes are so far only marginally unraveled. In fact, very little is known about which genes are activated or down-regulated to induce cell death.

Programmed Cell Death Phenotype in *Trypanosoma brucei*

Although the molecular mechanisms of PCD in *Trypanosoma brucei* remain to be solved, there are several reports that describe a clear cell death phenotype which shares characteristics

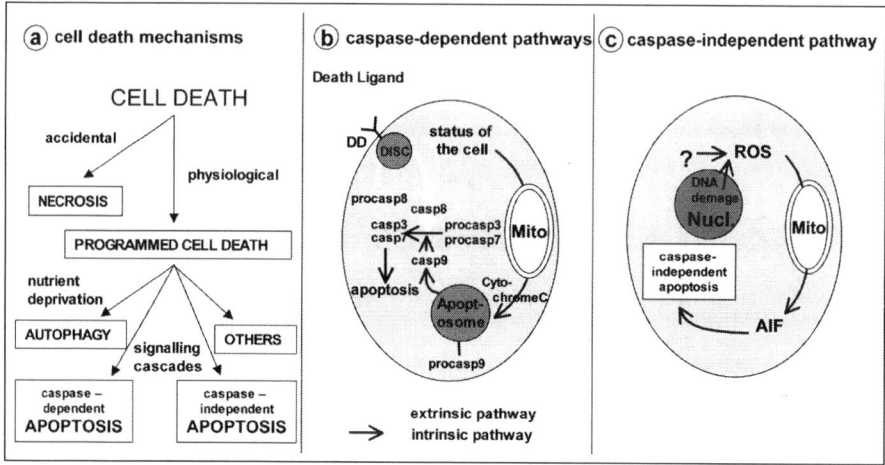

Figure 2. Cell death mechanisms in metazoa and protozoa. a) classification of cell death mechanisms. b) extrinsic and intrinsic pathways of caspase-dependent apoptosis in metazoa. c) caspase-independent apoptosis of metazoa and protozoa. (casp. = caspase, procasp = procaspase, Nucl. = nucleus; Mito = mitochondrion).

with apoptosis in higher eukaryotes.[14,18-23] However, classical apoptosis requires activation of caspases, which are not present in trypanosomes. They possess five metacaspases[24] and reports are available that metacaspases may be involved in PCD of other organisms,[24-28] but so far there are no evidences that metacaspases are involved in cell death in African trypanosomes.[29,30] Instead, it has been suggested that these proteins have a function in vesicle trafficking[30] and may be part of an interactome to regulate subcellular organisation.[29] Programmed cell death can occur in the complete absence of caspase activation, therefore alternative models of PCD including autophagy, paraptosis, mitotic catastrophe and the descriptive model of apoptosis-like and necrosis-like PCD have been proposed. In these cases, in addition to the mitochondrion, other organelles (such as lysosomes and the endoplasmic reticulum) may have an important function in the release and activation of death factors such as cathepsins, calpains and other proteases and nucleases.

Apoptosis-Like PCD (Caspase-Independent Apoptosis)

This type of PCD is morphologically different to apoptosis by showing less-compact chromatin condensation which gives more complex and lumpy shapes. Mechanistically, it is triggered by stress conditions, drugs, death receptors, etc., where molecules other than caspases, e.g., the apoptosis inducing factor (AIF), endonuclease G, cathepsins, or other proteases are involved.[31] In this sense, all phenotypes observed so far in procyclic or in bloodstream form trypanosomes can be classified as apoptosis-like PCD. So it was observed that *Trypanosoma brucei rhodesiense*, after transformation to the procyclic form in the fly, undergoes a cell death process[32] that displays a characteristic apoptotic morphology including surface membrane vesiculation and condensation of chromatin at the periphery of the nuclear membrane, while mitochondria remain intact.[18] These morphological features were also observed in procyclic parasites incubated in vitro with concanavalin A, where a differential expression of certain mRNAs was identified.[33] The concanavalin A receptor was found to be procyclin, the major cell surface glycoprotein in procyclic forms.[34] Incubation of procyclic parasites with xanthine oxidase induced DNA damage and cell death in a Ca^{2+}-dependent manner.[35] Likewise, in bloodstream forms of *T. b. gambiense* quercetin (3,3',4',5,7-pentahydroxyflavone), a potent immunomodulating flavenoid, was able to induce PCD as judged by detection of phosphatidylserine externalization using annexin V binding.[20] On the other hand, prostaglandin D_2 (PGD_2), a molecule shown to be primarily secreted by the stumpy form,[36] was reported to cause a cell death phenotype associated with loss of mitochondrial membrane potential, DNA degradation, chromatin condensation, cytoplasmic vacuolization and phosphatidylserine exposure[22] (Fig. 3a). Prostaglandins are molecules that possess a very short lifetime and carry out their actions by a paracrine or autocrine mechanism. Recently, it was shown that PGD_2 degradation to J-series metabolites enhanced the apoptosis-like cell death phenotype.[23] Interestingly, prostaglandin induced cell death was more prominent in stumpy than in slender parasites, since pre-incubation with a slender to stumpy differentiation inductor increased parasite's sensibility.[22] In addition, the induction of cell death in *T. brucei* by prostaglandins was demonstrated to be dependent on de novo protein synthesis and clearly associated with an increase in intracellular reactive oxygen species (ROS) concentration.[22,23] Using the ROS scavengers glutathione (GSH) or N-acetyl-L-cysteine (NAC), intracellular ROS generation was abolished as well as apoptosis-like features and cell death, indicating that ROS production plays a pivotal role in PCD in *T. brucei*. Moreover, it has been found that bloodstream trypanosomes cultured under high density condition show apoptosis-like features.[21] The authors demonstrated that a temperature shift from 37 °C to 27 °C protected trypanosomes against PCD, which was associated with an increase in expression of the trypanosome alternative oxidase (TAO). TAO is the unique terminal oxidase in bloodstream forms and responsible for removing excess reducing equivalents produced during glycolysis in the glycerol-3-phosphate DH reaction by transferring hydrogen to molecular oxygen.[37,38] Although procyclic parasites possess a functional mitochondrion, they also express TAO. It was observed that the inhibition of this enzyme in the procyclic form results in increased ROS production. Therefore, it was proposed that this enzyme can work as scavenger for mitochondrial superoxide.[38] Indeed, TAO over-expression in procyclic parasites contributed to the reduction in ROS production and to resistance of PCD in bloodstream forms.[21,39] Interestingly, incubation of bloodstream trypanosomes with prostaglandins and salicylhydroxamic acid (SHAM), an effective TAO inhibitor, had a synergistic effect on prostaglandin induced growth inhibition.[40] Recently, it was suggested that ROS can play a major role in killing trypanosomes entering the fly midgut, because addition of antioxidants to the tsetse blood meal was shown to significantly increase trypanosome midgut infection rates in *G. m. morsitans*.[41] Overall, these results indicate the involvement of ROS as mediators in the signalling pathway leading to PCD in *T. brucei*.

Autophagy

Type II autophagic or lysosomal cell death starts with the sequestration of cytoplasmic material, including organelles, by a multilayer membrane to form an autophagosome. The expansion of the lysosomal system followed by selective clearance of specific cell organelles by the autophagic vacuoles, plays an important role in cytoplasmic homeostasis.[42,43] However, an increased autophagy may lead to elimination of cells by self-digestion. Autophagy has been extensively studied in the yeast *Saccharomyces cerevisiae* and the molecular machinery that regulates the process is well known. In trypanosomes there is hardly any information about this mechanism. Recently, Rigden and collaborators published a genomic search in trypanosomatids for autophagy related genes.[44] 20 orthologues of yeast autophagy-implicated genes were found, which indicates that autophagy is genomically viable in these parasites as well. Indeed, experimental observations of trypanosomes incubated under starvation conditions show clear formation of autophagic vacuoles, as judged by monodansylcadaverine uptake (Fig. 3b). In addition, cells showing typical autophagic vacuoles have also been observed using transmission electron microscopy (Fig. 3a).

Central Role of ROS in Trypanosomal PCD

Reactive oxygen species (ROS) are generated as by-products of cellular metabolism, primarily in the mitochondria. When cellular production of ROS overwhelms the cells antioxidant capacity, damages to cellular macromolecules such as lipids, protein and DNA may ensue. To protect against the potentially damaging effects of ROS, cells possess several antioxidant enzymes such as superoxide dismutase (which reduces O_2^- to H_2O_2), catalase and glutathione peroxidase (which reduces H_2O_2 to H_2O). While superoxide dismutase is present in trypanosomes, catalase and

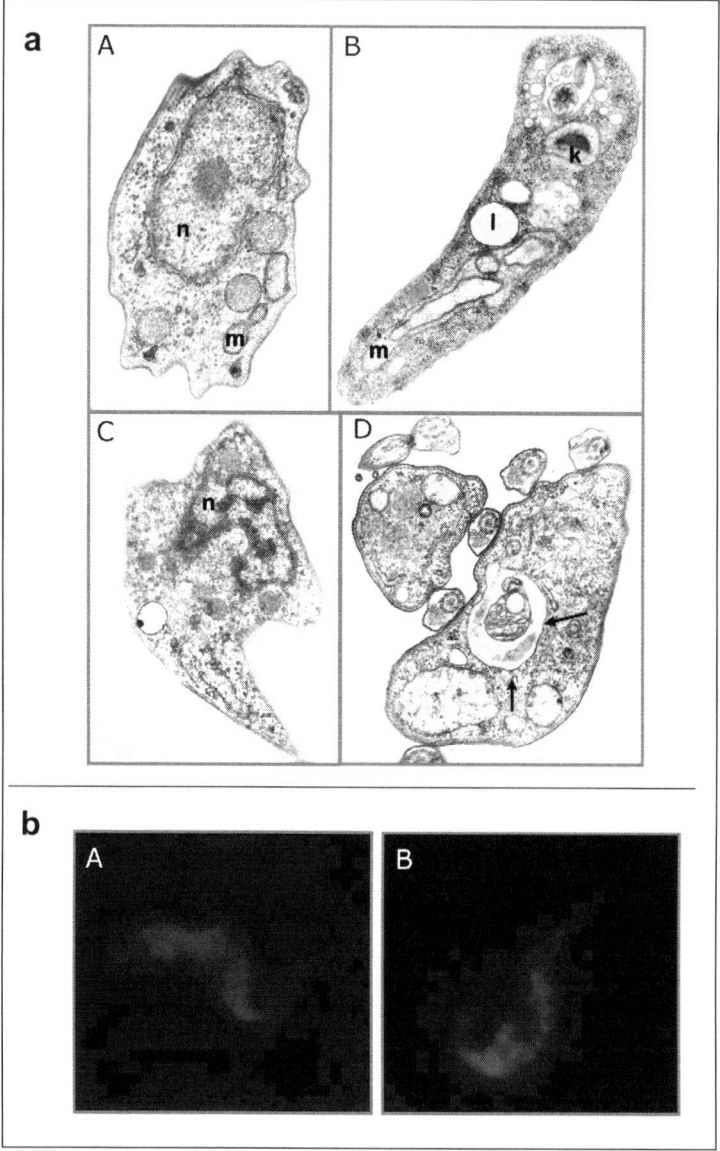

Figure 3. Evidences of PCD in bloodstream trypanosomes. a) Transmission electron microscopy of bloodstream trypanosomes showing morphological changes induced by prostaglandins (B and C) and typical autophagic vacuoles (D; see arrows), note the characteristic double membrane in the vacuole, wrapping cellular structures. Compare with the cytoplasm and organelles in the control cell (A). b) Trypanosomes stained with monodansylcadaverine (MDC). Cells were incubated for 3 h at 37°C, 5% CO_2, in a buffer containing 5 mM glucose plus an aminoacid cocktail (A) and without aminoacids (B) to induce starvation. The parasites were analysed using fluorescence microscopy. Trypanosomes under starvation condition displayed typical MDC positive vacuoles in their cytoplasm, a characteristic feature of autophagy. (n = nucleus; m = mitochondrium; l = lysosom; k = kynetoplast).

glutathione peroxidase are absent. Instead, trypanosomes possess a unique thiol metabolism in which the otherwise ubiquitous glutathione / glutathione reductase system is replaced by trypanothione and trypanothione reductase.[45] Trypanothione represents a key molecule to protect the parasites from oxidant damage and delivers reducing equivalents for DNA synthesis.[46,47] On the other hand, it was proposed that the trypanosome alternative oxidase (TAO) can also exert a role in the elimination of radicals arising within the mitochondrion.[38]

However, despite the well documented action of ROS as injurious by-products of cellular metabolism, ROS are also known as signaling molecules for physiological functions, such as regulation of the vascular tone, monitoring oxygen tension in the control of ventilation and erythropoietin production and signal transduction together with certain membrane receptors in various physiological processes.[48-50] In addition, ROS have been described to participate in the signal cascades associated to PCD.[51] Here it has been demonstrated that disturbance of the cellular redox equilibrium is involved in activation of PCD and that this represents the basis of many of the involved biological effects.[52-54]

Evidences accumulated so far indicate that ROS represent pivotal molecules in the induction and/or execution of PCD in African trypanosomes. Figure 4 summarizes interactions known so far regarding ROS and PCD in these parasites. As described above, compounds and conditions known to induce an apoptosis-like phenotype in trypanosomes, such as treatment with prostaglandin D_2 or its J-series metabolites, high density culture condition, treatment with xantine oxidase, or reduced survival of parasites within the tsetse flies midgut, were all associated with increased intracellular ROS concentration (Fig. 4). In all these cases use of ROS scavengers increased the resistance against PCD. Although, the mitochondrion represents the principal ROS source in most cells and since down-regulation of the mitochondrial enzyme TAO will lead to an increased ROS level and elevated cell death susceptibility, other ROS sources in trypanosomes can not be discarded. Oxidative stress was shown to change the intracellular Ca^{2+} distribution in trypanosomes by increased cytoplasmic and nuclear Ca^{2+} concentrations thus leading to endonuclease activation, DNA degradation and cell death[35] (Fig. 4). On the other hand, ROS increase could be associated with a decrease of DNA synthesis via trypanothione oxidation. The disulfide form of trypanothione (oxidized trypanothione) is a powerful inhibitor of the trypanedoxin reaction, which provides reducing equivalents for the synthesis of deoxyribonucleotides, essential for DNA synthesis[55] (Fig. 4). Therefore, the redox status of the cell may represent a check point for the regulation of desoxyribonucleotide synthesis and cellular proliferation. In summary, all these results confirm that ROS and the induction of oxidative stress in both, the procyclic and the bloodstream form of African trypanosomes seem to play a key role in PCD induction.

Physiological Relevance of PCD in *Trypanosoma brucei*

It has been proposed that trypanosomes could develop a suicide mechanism in order to maintain genetic stability within the population. In host-parasite interaction, parasite's PCD plays a critical role in the control of infection, since an uncontrolled increase of the parasite population would lead to the death of the host thus leading to elimination of the trypanosome clone. In fact, a strict population size control was shown to operate within tsetse flies infected with *Trypanosoma brucei rhodesiense*,[32] where (once established) the size of the parasite population remains constant for the lifetime of the fly. This balance is reached by an equilibrium between parasite multiplication and a regulated cell death. This fact is important, because the parasites and the fly compete for proline as energy source. In the case of the mammalian host, it has been postulated that the elimination of parasites by PCD could avoid an exacerbated immunological reaction. It has been observed that parasites cultured in vitro at high density show apoptosis-like features, which, if occurring in vivo during natural infection, may contribute to control the population density in the mammalian host. In addition, differentiation from slender to stumpy parasites was demonstrated to be a mechanism to control parasitaemia, since stumpy forms are arrested cells, unable to proliferate. Interestingly, the stumpy population can be self controlled since this form of the parasite has an elevated capacity to produce and secrete prostaglandin D_2, which induces apoptosis-like PCD mainly on stumpy

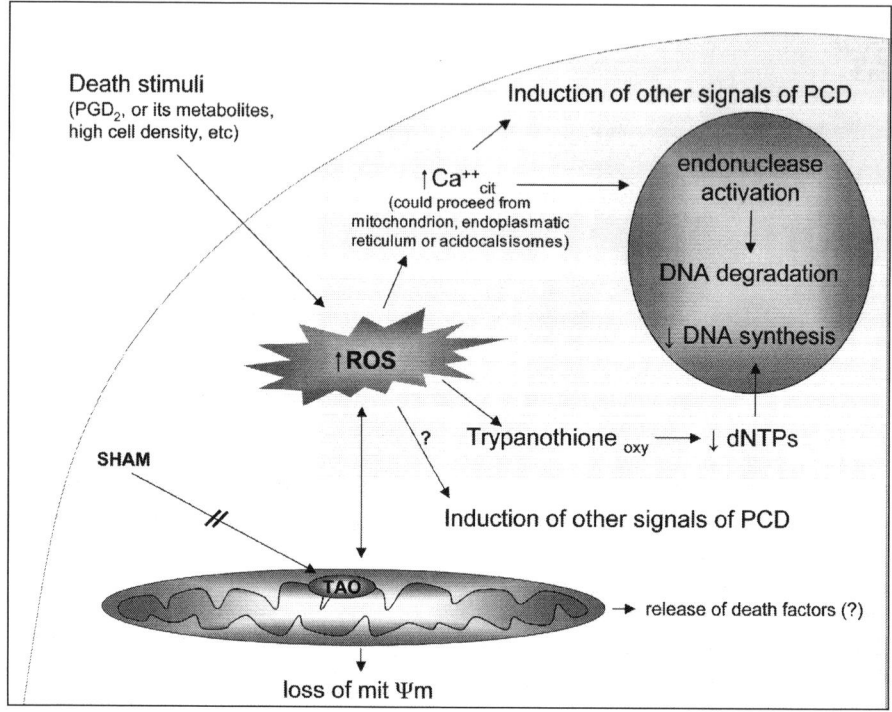

Figure 4. Schematic representation of the central role of ROS in the induction and/or execution of programmed cell death in *Trypanosoma brucei*. Events known to induce PCD in African trypanosomes were shown to increase intracellular ROS. Experimental evidences indicate that ROS can induce different responses which lead to cell death.

parasites.[22] Therefore, this mechanism guarantees persistence of the infection within the host. Acquisition of any form of PCD in trypanosomes represents a considerable advantage as it allows self-regulation of the population density independent from the host, thus assuring perpetuation of the trypanosome species during evolution.

Acknowledgements

This work was supported by the Deutsche Forschungsgemeinschaft (MD) and the Wellcome Trust (SCW). KF and NU are recipients of a DAAD and CDCH Scholarships, respectively.

References

1. Heurlier K, Denervaud V, Haas D. Impact of quorum sensing on fitness of Pseudomonas aeruginosa. Int J Med Microbiol 2006; 296:93-102.
2. Duszenko M, Figarella K, Macleod ET et al. Death of a trypanosome: a selfish altruism. Trends in Parasitol 2006; 22:536-542.
3. Clayton CE and Michels P. Metabolic compartmentation in African trypanosomes. Parasitol Today 1996; 12:465-471.
4. Barry JD, McCulloch R. Antigenic variation in trypanosomes: enhanced phenotypic variation in a eukaryotic parasite. Adv Parasitol 2001; 49:1-70.
5. Turner CM, Aslam N, Dye C. Replication, differentiation, growth and the virulence of Trypanosoma brucei infections. Parasitology 1995; 111:289-300.
6. Tyler KM. Maintenance of parasitaemia—is it to die for? Kinetoplastid Biol Dis 2003; 2:2.
7 Borst P, Rudenko G, Taylor MC et al. Antigenic variation in trypanosomes. Arch Med Res 1996; 27:379-388.

8. Hesse F, Selzer PM, Muhlstadt K et al. A novel cultivation technique for long-term maintenance of bloodstream form trypanosomes in vitro. Mol Biochem Parasitol 1995; 70:157-166.

9. Vassella E, Reuner B, Yutzy B et al. Differentiation of African trypanosomes is controlled by a density sensing mechanism which signals cell cycle arrest via the cAMP pathway. J Cell Sci 1997; 110: 2661-2671.

10. Van Houdt R, Moons P, Hueso Buj M et al. N-acyl-L-homoserine lactone quorum sensing controls butanediol fermentation in Serratia plymuthica RVH1 and Serratia marcescens MG1. J Bacteriol 2006; 188:4570-4572.

11. Saito T, Taylor GW, Yang JC et al. Identification of new differentiation inducing factors from Dictyostelium discoideum. Biochim Biophys Acta 2006; 1760:754-761.

12. Billker O, Lindo V, Panico M et al. Identification of xanthurenic acid as the putative inducer of malaria development in the mosquito. Nature 1998; 392:289-292.

13. Raff MC. Social controls on cell survival and cell death. Nature 1992; 356:397-400.

14. Debrabant A, Lee N, Bertholet S et al. Programmed cell death in trypanosomatids and other unicellular organisms. Int J Parasitol 2003; 33:257-267.

15. Lewis K. Programmed death in bacteria. Microbiol Mol Biol Rev 2000; 64:503-514.

16. Tan KS and Nasirudeen AM. Protozoan programmed cell death--insights from Blastocystis deathstyles. Trends Parasitol 2005; 21:547-550.

17. Szallies A, Merkel P, Mielenz H et al. A role for yeast metacaspase 1 in ubiquitination: Genetic interactions with RSP5 and other genes require a conserved tyrosine residue. 2007; submitted

18. Welburn SC, Dale C, Ellis D et al. Apoptosis in procyclic Trypanosoma brucei rhodesiense in vitro. Cell Death Differ 1996; 3:229-236.

19. Nguewa PA, Fuertes MA, Valladares B et al. Programmed cell death in trypanosomatids: a way to maximize their biological fitness? Trends Parasitol 2004; 20:375-380.

20. Mamani-Matsuda M, Rambert J, Malvy D et al. Quercetin induces apoptosis of Trypanosoma brucei gambiense and decreases the proinflammatory response of human macrophages. Antimicrob Agents Chemother 2004; 48:924-929.

21. Tsuda A, Witola WH, Ohashi K et al. Expression of alternative oxidase inhibits programmed cell death-like phenomenon in bloodstream form of Trypanosoma brucei rhodesiense. Parasitol Int 2005; 54:243-251.

22. Figarella K, Rawer M, Uzcategui NL et al. Prostaglandin D2 induces programmed cell death in Trypanosoma brucei bloodstream form. Cell Death Differ 2005; 12:335-346.

23. Figarella K, Uzcategui NL, Beck A et al. Prostaglandin-induced programmed cell death in Trypanosoma brucei involves oxidative stress. Cell Death Differ; 2006; 13:1802-1814.

24. Szallies A, Kubata BK and Duszenko M. A metacaspase of Trypanosoma brucei causes loss of respiration competence and clonal death in the yeast Saccharomyces cerevisiae. FEBS Lett 2002; 517:144-150.

25. Kosec G, Alvarez VE, Aguero F et al. Metacaspases of Trypanosoma cruzi: possible candidates for programmed cell death mediators. Mol Biochem Parasitol 2006; 145:18-28.

26. Mazzoni C, Herker E, Palermo V et al. Yeast caspase 1 links messenger RNA stability to apoptosis in yeast. EMBO Rep 2005; 6:1076-1081.

27. Vercammen D Type II metacaspases Atmc4 and Atmc9 of Arabidopsis thaliana cleave substrates after arginine and lysine. J Biol Chem 2004; 279:45329-45336.

28. Watanabe N, Lam E. Two Arabidopsis metacaspases AtMCP1b and AtMCP2b are arginine/lysine-specific cysteine proteases and activate apoptosis-like cell death in yeast. J Biol Chem 2005; 280:14691-14699

29. Szallies A. [PhD Thesis]. Germany: University of Tubingen; 2004.

30. Helms MJ, Ambit A, Appleton P et al. Bloodstream form Trypanosoma brucei depend upon multiple metacaspases associated with RAB11-positive endosomes. J Cell Sci 2006; 119:1105-1117.

31. Jaattela M, Tschopp J. Caspase-independent cell death in T-lymphocytes. Cell 2003; 114:181-190.

32. Maudlin I, Welburn SC. Maturation of trypanosome infections in tsetse. Exp Parasitol 1994; 79:202-205.

33. Murphy NB, Welburn SC. Programmed cell death in procyclic Trypanosoma brucei rhodesiense is associated with differential expression of mRNAs. Cell Death Differ 1997; 4:365-470.

34. Pearson T, Beecroft R, Welburn SC et al. The major cell surface glycoprotein procyclin is a receptor for induction of a novel form of cell death in African trypanosomes in vitro. Mol Biochem Parasitol 2000; 111:333-349.

35. Ridgley EL, Xiong ZH, Ruben L. Reactive oxygen species activate a Ca^{2+}-dependent cell death pathway in the unicellular organism Trypanosoma brucei brucei. Biochem J 1999; 340:33-40.

36. Kubata BK, Duszenko M, Kabututu Z et al. Identification of a novel prostaglandin f(2alpha) synthase in Trypanosoma brucei. J Exp Med 2000; 192:1327-1338.

37. Chaudhuri M, Ajayi W, Hill GC. Biochemical and molecular properties of the Trypanosoma brucei alternative oxidase. Mol Biochem Parasitol 1998; 95:53-68.

38. Fang J, Beattie DS. Alternative oxidase present in procyclic Trypanosoma brucei may act to lower the mitochondrial production of superoxide. Arch Biochem Biophys 2003; 414:294-302.

39. Tsuda A, Witola WH, Konnai S et al. The effect of TAO expression on PCD-like phenomenon development and drug resistance in Trypanosoma brucei. Parasitol Int 2006; 55:135-142.
40. Figarella K 2005 [PhD Thesis] Germany: University of Tubingen.
41. Macleod ET. 2005 [PhD Thesis]: University of Edinburgh, Scotland.
42. Ogier-Denis E, Codogno P. Autophagy: a barrier or an adaptive response to cancer. Biochim Biophys Acta 2003; 1603:113-128.
43. Wang CW, Klionsky DJ. The molecular mechanism of autophagy. Mol Med 2003; 9:65-76.
44. Rigden DJ, Herman M, Gillies S et al. Implications of a genomic search for autophagy-related genes in trypanosomatids. Biochem Soc Trans 2005; 33:972-974.
45. Fairlamb AH, Cerami A. Metabolism and functions of trypanothione in the Kinetoplastida. Annu Rev Microbiol 1992; 46:695-729.
46. Krauth-Siegel RL, Coombs GH. Enzymes of parasite thiol metabolism as drug targets. Parasitol Today 1999; 15:404-409.
47. Krauth-Siegel RL, Schmidt H. Trypanothione and tryparedoxin in ribonucleotide reduction. Methods Enzymol 2002; 347:259-266.
48. Finkel T. Oxygen radicals and signaling. Curr Opin Cell Biol 1998; 10:248-253.
49. Rhee SG. Redox signaling: hydrogen peroxide as intracellular messenger. Exp Mol Med 1999; 31:53-59.
50. Droege W. Free radicals in the physiological control of cell function. Physiol Rev 2002; 82:47-95.
51. Thannickal VJ, Fanburg BL. Reactive oxygen species in cell signaling. Am J Physiol Lung Cell Mol Physiol 2000; 279:1005-1028.
52. Russo T, Zambrano N, Esposito F et al. A p53-independent pathway for activation of WAF1/CIP1 expression following oxidative stress. J Biol Chem 1995; 270:29386-29391.
53. Esposito F, Cuccovillo F, Vanoni M et al. Redox-mediated regulation of p21waf/cip1 expression involves a posttranscriptional mechanism and activation of the mitogen-activated protein kinase pathway. Eur J Biochem 1997; 245:730-737.
54. Chen YC, Shen SC, Tsai SH. Prostaglandin D2 and J2 induce apoptosis in human leukemia cells via activation of caspase 3 cascade and production of reactive oxygen species. Biochim Biophys Acta 2005; 1743:291-304.
55. Dormeyer M, Reckenfelderbaumer N, Ludemann H et al. Trypanothione-dependent synthesis f deoxyribonucleotides by Trypanosoma brucei ribonucleotide reductase. J Biol Chem 2001; 276:10602-10606.

CHAPTER 5

Molecular Analysis of Programmed Cell Death by DNA Topoisomerase Inhibitors in Kinetoplastid Parasite *Leishmania*

Nilkantha Sen, Bijoylaxmi Banerjee and Hemanta K. Majumder*

Abstract

DNA topoisomerases of kinetoplastid parasites represent a family of DNA processing enzymes that essentially solve the topological problems not only in nuclear DNA but also in kinetoplastid DNA. Due to their indispensable function in cell biology they are valuable potential targets for many antileishmanial drugs to induce programmed cell death (PCD). PCD is thought to have evolved to regulate growth and development of multicellular organisms. Kinetoplastid organisms have developed an altruistic mechanism to promote and maintain the clonality within their population. Characterization of PCD by topoisomerase inhibitors could provide information regarding their pathogenesis which could be exploited to develop new targets to limit their growth and treat the disease they cause.

DNA topology affects fundamental processes of life and DNA topoisomerases constitute a growing family of nuclear enzymes that regulate unconstrained DNA supercoils in all living species, bacteria, archea, kinetoplastids and eukaryotes. Changes in DNA topology are required for virtually all DNA dependent events such as DNA replication, transcription, recombination, repair, nucleosome remodeling, chromosome condensation and segregation.[1,2] Topoisomerases are divided into two classes, based primarily on their mode of cleaving DNA. Type I DNA topoisomerases act by making a transient nick on a single strand of duplex DNA, passing another strand through the nick and changing the linking number by one unit.[3] Type II topoisomerases act by transiently nicking both strands of the DNA, passing another double stranded DNA segment through the gap and changing the linking number by two with the help of ATP molecules.[4]

DNA Topoisomerases in Kinetoplastid Parasites—Why So Unique?

DNA Topoisomerase I

Recently, the emergence of the bi-subunit topoisomerase I of *Leishmania*[5,6] in the kinetoplastid family have brought a new dimension in topoisomerase research related to evolution and functional conservation of type IB family. The core DNA binding domain, LdTOP1L (73 kDa) and the catalytic domain LdTOP1S, (29 kDa) harboring the consensus SKXXY motif lie in separate

*Corresponding Author: Hemanta K. Majumder—Division of Molecular Parasitology, Indian Institute of Chemical Biology. 4, Raja S.C Mullick Road, Kolkata 700 032, India. Email: hkmajumder@iicb.res.in

Programmed Cell Death in Protozoa, edited by José Manuel Pérez Martín.

sub-units. This unusual structure of DNA topoisomerase I may provide a missing link in the evolution of type IB enzyme. The reconstituted enzyme (LdTOP1LS) is characterized by a direct 1:1 molar interaction between the large (LdTOP1L) and the small (LdTOP1S) subunits. A precise insight into the intra cellular location of the expressed bi- subunit topoisomerase I proteins in *L. donovani* revealed that topoisomerase I was localized both in nucleus and kinetoplast.[5] Moreover, it was also established that, 1-39 amino acid residues of the large subunit (LdTOP1L) of the unusual bi-subunit enzyme regulate DNA dynamics during relaxation by controlling noncovalent DNA binding or by coordinating DNA contacts by other parts of the enzyme and CPT sensitivity. The residues with in 40-99 amino acid region of LdTOP1L appear to be important in relation to interaction with LdTOP1S.[7] The multimeric structure of topoisomerase IB in kinetoplastid parasites provides another vivid example of the unusual nucleic acid metabolism in these pathogens. Although the type IB catalytic activity in trypanosomatids is entirely conventional in terms of its substrate preference, Mg^{++} independence and CPT and flavone sensitivity,[8] the reason for the existence of heterodimeric subunit organization still remains to be determined. Comparing amino acid sequences of trypanosomatid topoisomerase IB with monomeric topoisomerase IB reveals that each subunit contributes essential and highly conserved catalytic residues that characterize both the type IB topoisomerases and tyrosine recombinases. This has led to the concept that these two enzyme classes share a common ancestral catalytic domain that formed over time with different N-terminal domains.[9] So, kinetoplastids may provide missing link in evolution between the hypothesized ancient independent catalytic domain and the contemporary fused constructs. Moreover, multiple classical bipartite nuclear localization signals are found in the large but not the small subunit enzyme. So the complex assembles in the cytosol before nuclear importation. Neither subunit contains a detectable mitochondrial targeting sequence but their localization in kinetoplast is an interesting finding.

DNA Topoisomerase II

Genes for topoisomerase II have been isolated and sequenced from kinetoplastid organisms, *Crithidia fasciculate*.[10] *L. donovani*,[11] *Trypanosoma brucei*,[12] *Trypanosoma cruzi*[13] and *Bodo saltans*.[14] Although smaller than the genes reported from other eukaryotes, they share the expected functional domains and are more homologous to eukaryotic than prokaryotic type II enzymes with 30-35% identity and 45-65% similarity to human topoisomerase II. The predicted proteins are 137-160 kDa in size. The recombinant *L. donovani* enzyme showed ATP-dependent decatenating activity and could complement a temperature sensitive topoisomerase II mutant of *S. cerevisiae*.[15] A precise insight into the intra cellular location of the expressed topoisomerase II protein in *L. donovani* was gained by indirect immunoflourescence. Polyclonal antisera raised against a conserved portion of LdTOP2 showed nuclear and kinetoplast localizations for *L. donovani* topoisomerase II protein.[16]

Topoisomerase II is a highly potential drug target because it has an indispensable function in cell biology and it lacks biological redundancy. So far the study made on LdTOP2 protein enhances our understanding in the following ways: (1) there is a clear distinction between the regions of LdTOP2 required for ATP hydrolysis and trans-esterification reaction, (2) the catalytic domain per se is capable of recognizing the target sequence for transesterification reaction and (3) ATP hydrolysis which is crucial for topoisomerase II function in vivo is not required for DNA binding or cleavage. It was also observed that the core domain of LdTOP2 is involved in the formation of a covalent complex with DNA with Tyr 775 being the active site residue. Moreover it was also established that the ATPase, DNA cleavage and the C-terminal domain of LdTOP2 are relatively independent in topoisomerase structure, with the entire domain working in conjunction for full topoisomerase activity. A future aspect of this work lies in exploiting the differences between the CTD of the host and parasite enzyme by integrating structural information with biochemical experimentations, so as to delineate the common and distinguishing feature of the host and parasite enzyme.[17]

Role of DNA Topoisomerases in Kinetoplastids

In the last two decades, considerable attention has been devoted to the topoisomerases of kinetoplastid parasites, namely *T. brucei, T. cruzi, C. fasciculata* and *L. donovani*. Such an importance can be attributed to i) the emergence of the unicellular kinetoplastid parasites as pathogenic organisms, second only to the worldwide menace of malaria caused by *Plasmodium* species and ii) the uniqueness and novelty of mitochondrial DNA (kDNA) of the kinetoplastid parasites.

The kDNA shows one of the most amazing topological organizations of any DNA found in nature. It consists of two classes of circular molecules of different sizes, the maxicircle and minicircles. However, the huge structure of kDNA network possesses two significant problems. The first problem is topological and questions the ability of an interlocked DNA circle to replicate. The second problem concerns book keeping, questioning the ability of a parasite to ensure that each minicircle replicates only once per cell cycle.

The current model of replication answers both the above questions. First, the minicircles are released randomly from the center of the network by topoisomerase II. These free minicircles undergo replication as θ structures, as would any DNA circle in any other cell.[18] Therefore, the whole network does not decatenate completely into free minicircles. Instead, at any given point of time during the S phase, there are several hundred free minicircles that undergo replication.[19] The progeny minicircles, containing nicks and gaps, are subsequently attached to the network periphery. These nicks and gaps distinguish replicated minicircles from the un-replicated ones. Only when all the minicircles have been replicated, the nicks get repaired. With ongoing replication, the total number of minicircles increases and as a result the central zone shrinks and the peripheral zone enlarges.

The process of maxi circle replication is very similar to that of the above mentioned mini circle replication with only one exception. Maxi circles replicate when they are still linked to the network. Their replication initiates in a unique region containing the sequence GGGGGTTGGTGT, which is one base shorter than the mini circle replication origin[20,21] and the replication fork moves uni directionally as θ structures. These data therefore contrast an earlier report that suggested maxi circles replicate by a rolling circle mechanism.[22]

Following the completion of mini circle and maxi circle replication, the network undergoes an extensive remodeling. As a result of mitochondrial membrane enlargement the available network space also increases and accordingly the network expands. During this time, the surface area of the network becomes double that of the pre replicated form.[23] Finally, topoisomerase mediated unlinking of neighboring circles between two daughter cells result in splitting. The characteristic oval shape is subsequently established by the remodeling of the flat-edged daughter networks.

DNA Topoisomerases as a Potential Target for Apoptosis

Beyond the essential cellular functions, eukaryotic topoisomerases I and II and bacterial DNA gyrases are the targets for a number of clinically important drugs. The known DNA topoisomerase drugs can be divided into two classes. The class I drugs can act by stabilizing the covalent topoisomerases-DNA complexes and are also referred to as "topoisomerase poison".[24] They include the bacterial gyrase inhibitors e.g quinolones, the DNA topoisomerase I drug camptothecin[25,26] and the DNA topoisomerase II drugs doxorubicin, amsacrine, etoposide, teniposide, quercetin and related flavonoids.[27]

The class II drugs interfere with the catalytic functions of DNA topoisomerases without trapping the covalent complexes and are referred to as "catalytic inhibitors". This class of drugs includes the coumermycin family of antibiotics that act on bacterial gyrase and the eukaryotic topoisomerase II inhibitors suramin, fostriecin, merbarone and bis (2,6-dioxopiparazine).[28,29]

Topoisomerase cleavage complexes can be further converted into DNA lesions during DNA replication and transcription. In replicating cells, topoisomerase I cleavage complexes can initiate DNA damage and cell death once a replication fork collides with a stabilized topoisomerase I cleavage complexes.[30] The resulting lesion is a replication-mediated DNA double strand break associated with a topoisomerase I covalent crosslink. It is also likely that cell death can be initiated by collisions of transcription complexes with the stalled topoisomerase I cleavage complexes. Transcription coupled cell death seems to be particularly relevant for nonreplicating cells such as neurons and lymphocytes. The cellular effects of topoisomerase II cleavage complexes and the pathways implicated in their repair are less known.

Significance of Apoptosis in Kinetoplastid Parasites

Programmed cell death (PCD) is a mechanism involving differential expression of specific target genes and apoptosis is the most common phenotype of PCD.[31] Apoptotic cells are characterized by morphological markers such as cell shrinkage, PS exposure, membrane blebbing, DNA fragmentation and packaging of cell contents into apoptotic bodies. This kind of cell death is non-inflammatory, especially because it prevents the release of cytoplasmic contents, since apoptotic bodies are quickly cleared by phagocytes. Here we discuss about what are the benefits of apoptosis for unicellular organisms. Two main hypotheses have been proposed to answer this question. First, cell death can be important for population size control when there are insufficient nutrients or to avoid host cell death. In this case unicellular apoptotic cells show an altruistic behavior, dying for the benefit of others. The second explanation is that apoptotic cells, which will not necessarily die, could provide signals that enhance the survival of the entire population.

It was previously shown that amastigote forms of *L. amazonesis* expose PS on their surface and use this ligand to penetrate host macrophages. PS recognition leads to enhanced TGF-β secretion and interleukin10 (IL-10) mRNA production. Pretreatment of amastigotes with annexin V-which binds to PS at high calcium concentrations inhibited amastigote infectivity by at least 50% and significantly reduced TGF-β1 production by infected macrophages. Thus PS signaling in amastigote forms functions like apoptotic cells according to a mechanism denoted by Wanderley et al as "apoptotic mimicry".[32]

The signals initiating PCD include growth arrest either due to nutritional deprivation or by drugs, inhibitors of signaling molecules, oxidative stress, heat shock, aging, differentiations and mutations in cell cycle regulating genes. Some of these signals were shown to initiate a cascade of events such as loss of mitochondrial membrane potential followed by induction of caspases like activity, induction of inter nucleosomal DNA cleavage and changes in plasma membrane permeability. The ultimate end results of these processes are nuclear condensation and DNA ladder formation which suggest that the type of PCD pathway observed in unicellular organisms may have some similarities to the PCD process in multicellular organisms.[33]

Recently, meta caspases, a group of cysteine proteases that are present in unicellular eukaryotes and that belong to the caspase-para csapase-meta caspase super family have been shown to have a role in protozoan PCD.[34] Indeed recent evidences indicate that Sir2 proteins (member of histone deacetylase) are involved in the regulation of p53–dependent apoptotic response via deacetylation of p53 proteins.[35] Furthermore, recent investigations have shown that changes in the level of elongation factor (EF-1α) may be one of the pivotal factors regulating the rate of apoptosis.[36] Indeed an anti-sense EF-1α provides cell protection against death upon starvation whereas EF-1α over expression leads to a faster rate of cell death.

DNA Topoisomerase Mediated Apoptosis in Kinetoplastids Significantly Differs from Mammalian Cells

Before going in to the details of topoisomerase mediated apoptosis in kinetoplastids we will discuss very briefly about the main signaling pathway from topoisomerase mediated DNA damage to apoptosis in mammalian cells. The pathway is schematically represented in Figure 1.

The apoptotic mitochondrial pathway: Like many other apoptotic stimuli, topoisomerase inhibitors induce mitochondrial membrane permeabilization,[37] a process under the control of BCl-2 related protein. The BCl-2 family members include antiapoptotic proteins (e.g., BCl-2, BCl-XL, MCl-I and BCl-W) and proapoptotic proteins (e.g., Bax, Bak, BCl-XS, Bad, Bid, Bik) that prevent or promote mitochondrial membrane permeabilization respectively. Permeabilization of the outer mitochondrial membrane induces the leakage of proapoptotic proteins normally confined to the

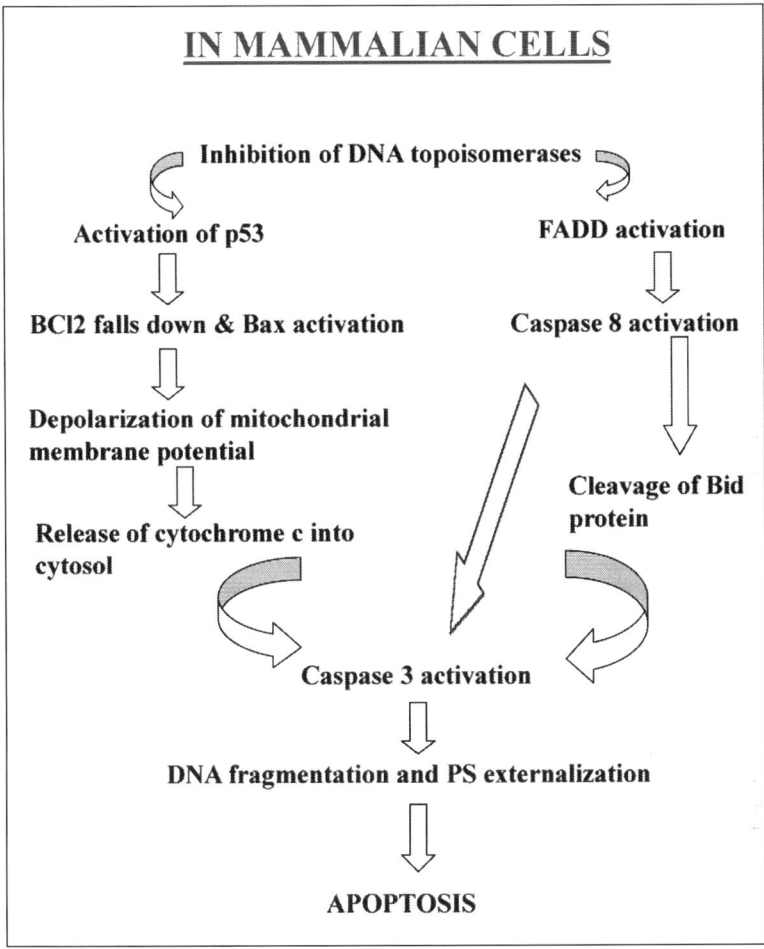

Figure 1. Possible mechanism of apoptotic signalling pathway by DNA topoismerase inhibitors.

intermembrane space of the mitochondria. These pro apoptotic proteins include cytochrome c, Smac/DIABLO, AIF (Apoptosis inducing factor) and endonuclease G.[38]

AIF and Endoclease G directly induce nuclear modifications, whereas cytochrome c, Smac/DIABLO promote caspases activation. In the cytosol, cytochrome c induces oligomerization of Apaf 1 in the presence of ATP and dATP. Apaf1 oligomers recruit procaspase 9 molecules in a complex called the apoptosome. Formation of apoptosome causes activation of caspases 9 from procaspase 9 which subsequently activate caspases 3. Some of heat shock proteins such as Hsp 60 and Hsp 10 are also released from mitochondria and facilitate activation of procaspase 3.

The death Receptor-dependent pathway: At least in some cell types, apoptotic response to topoisomerase inhibitors involve the death receptor Fas, also known as APO 1 or CD95. Fas is a 45 kDa, type I transmembrane receptor that belongs to the TNF receptor superfamily.[39] Crosslinking of Fas by its natural ligand (Fas L) induces clustering of Fas, which in turn recruits the cytosolic FADD and procaspase 8 to form a complex known as the death inducing signal complex (DISC). Oligomerization of procaspase 8 in the DISC results in the autoactivation of mature caspases 8.[40] In some cell types activated caspases 8 formed at the DISC induce direct cleavage of procaspase 3 independently of mitochondria. In other cell types, caspase 8 cleaves Bid protein generating an active fragment that activates the mitochondrial death pathway. The detection of topoisomerase mediated DNA lesions is ensured by large proteins (and protein complexes), commonly called DNA sensors, that bind to DNA breaks and activate checkpoint and repair proteins. DNA-PK, ATM and ATR or serine/threonine kinases from the PI-3 kinase family are known to be activated by topoisomerase mediated DNA damage. ATM, DNA-PK and ATR mediate their proapoptotic effects by phosphorylating a large number of substrates including several kinases such as c-Abl and the checkpoint protein chk 2, which themselves can activate several pathways leading to apoptosis.

Apoptosis in Kinetoplastid Parasites by DNA Topoisomerase Inhibitors

From our laboratory it was established for the first time that inhibition of DNA topoisomerase by nonoxidant CPT induces oxidative stress in leishmanial cells unlike other mammalian cells. The important role of ROS in the regulation of the apoptosis like death in parasites may indicate the origin and the primary purpose of the suicide process. ROS are the byproducts of respiration and occur in every aerobic organism. ROS are highly reactive and modify proteins, lipids and nucleic acids. So ROS mediated cell damage is a frequent event in apoptosis. CPT induced oxidative stress causes depletion of intracellular GSH level as well as increase in the level of lipid per oxidation.[41] It was also shown that CPT induced oxidative stress and lipid per oxidation increase the level of cytosolic Ca^{++} through both intracellular and extra cellular sources. Loss of mitochondrial membrane potential is an irreversible event in apoptosis. CPT induced oxidative stress and increase in cytosolic Ca^{++} level were responsible for the loss of $\Delta\Psi_m$ in leishmanial cells. Intracellular K^+ is known to be one of the most important determinants for the maintenance of ionic balance, which is directly related to osmotic pressure inside cells.[42] Most of the cells can maintain and achieve the osmotic balance through the continuous activity of an ATP dependent Na^+-K^+ pump that exchanges $3Na^+$ for $2K^+$ against the electrochemical gradient. Inhibition of topoisomerases by CPT causes impairment of Na^+-K^+ ATPase pump, which results in subsequent decrease in intracellular K^+ level and facilitates apoptosis in leishmanial cells. Moreover the release of cytochrome c into the cytosol after the loss of $\Delta\Psi_m$ and activation of caspase like proteases involved in apoptosis provide substantial evidences in support of the fact that PCD machinery in unicellular organisms like kinetoplastids may have evolved through the horizontal gene transfer or evolutionary convergence. In vitro study of the activation of caspase like proteases in different pH and KCl buffer revealed that lower pH accelerates the rate of formation of active apoptotic complex as opposed to increasing the V_{max} of cytochrome c and dATP activated caspase 3 like proteases as well as apoptotic nucleases.[42] This suggests that physiological concentration of K^+ has a direct inhibitory effect on the apoptotic nucleases and a decrease in intracellular K^+ concentration is a prerequisite for apoptotic DNA degradation.

Polyphenols including luteolin are common components which naturally exist in plants and have been demonstrated to show several biological properties including free radical scavenging, anti inflammatory and anti carcinogenesis activities. Unlike CPT, luteolin is a potent inhibitor of both DNA topoisomerase I and II in vitro and in vivo. From our laboratory it was previously established that luteolin and quercetin inhibited the growth of *L. donovani* promastigotes and axenic amastigotes, inhibited DNA synthesis in promastigotes and promoted topoisomerase II mediated linearization of kDNA minicircles. These compounds arrested cell cycle in promastigotes leading to apoptosis as indicated by PI permeability and binding of annexin V to the cell surface.[43] So the elucidation of effector signaling pathway after inhibition of DNA topoisomerase I and II in leishmanial cells remain unexplored. As DNA topoisomerase II controls replication of kinetoplastid as well as nuclear DNA in leishmanial cells, inhibition of DNA topoisomerases by luteolin causes loss of mitochondrial DNA and leads to formation of the dyskinetoplastid cells. The loss of mitochondrial DNA causes reduction in the activities of complex I, II, III and IV of electron transport chain although the ATPase activity of complex V remains unaltered. The inactivation of ETC is associated with decrease in mitochondrial as well as glycolytic ATP production, which is responsible for loss of $\Delta\Psi_m$ and alteration of mitochondrial structure and propagates apoptosis like death through caspase 3 like protease activation. Apoptosis like death in kinetoplastid parasites is schematically represented in Figure 2. Collectively, this study has provided substantial evidences in support of the fact that dyskinetoplastidy in leishmanial cells leads to apoptosis through mitochondria dependent pathway as well as clarified the importance of glycolytic ATP in apoptosis like death in mt-DNA depleted cells after inhibition of DNA topoisomerase inside leishmanial cells.[44] Moreover, it was demonstrated that the arsenite resistant strain to be cross resistant to novobiocin (catalytic inhibitor of DNA topoisomerase II), which induces dose and time dependent increase in apoptotic cell death in both the strains (Ld wt and LdAs20) of *L. donovani*. This has been correlated to over expression of topoisomerase II and was substantiated by differential inhibition of enzyme activity in wild type and arsenite resistant *L. donovani* cells.[45]

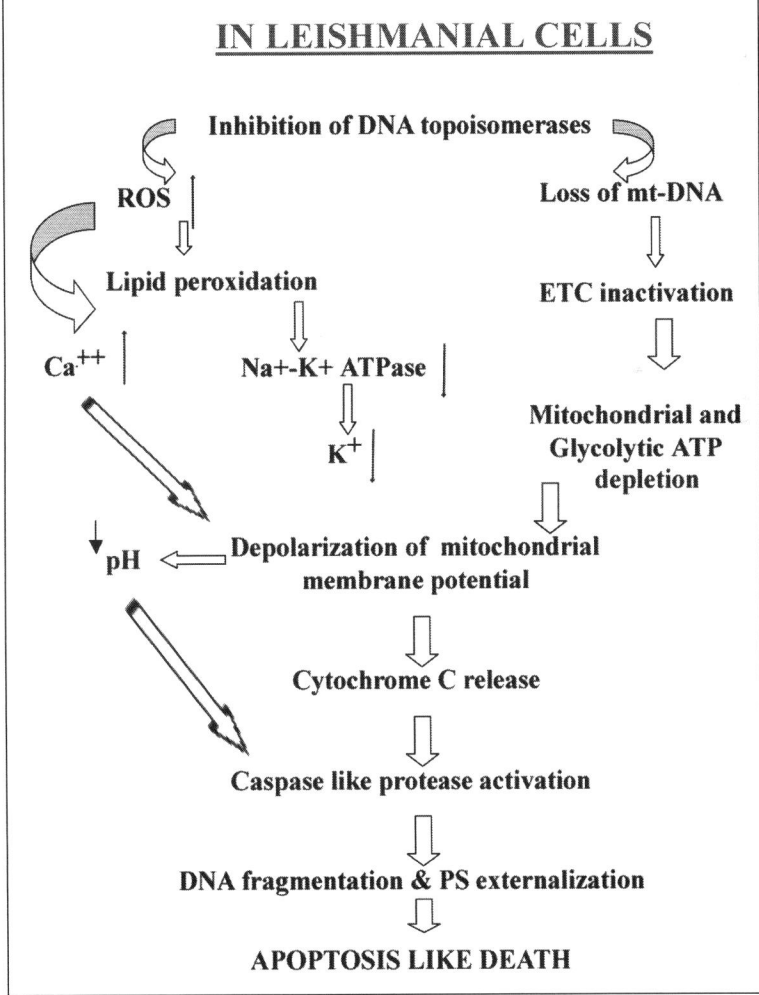

Figure 2. Possible mechanism of apoptotic signalling pathway by DNA topoisomerase inhibitors.

Concluding Remarks

It has been suggested that PCD is an ancient mechanism of self-destruction that existed before the advent of multicellularity. Studies of parasitic protozoa have shown that DNA topoisomerase mediated PCD pathways in these organisms are both conserved and unique properties. Moreover, oxidative stress mediated cell death is induced by topoisomerase inhibitors confirming the evolutionarily ancient link between ROS and cell suicide. The classical effector molecules that regulate PCD of mammalian cells induced by topoisomerase inhibitors, are absent in leishmanial cells. So it is important to understand more clearly the interaction between topoisomerase with chromatin inside cells in the presence of their inhibitors. This may help us to clarify the unicellular apoptotic pathway induced by topoisomerase inhibition. Acetylation of core histones has been shown to weaken the histone—DNA interaction and consequently increases the accessibility of topoisomerases to DNA. Deacetylation of histones is a known mechanism for chromatin condensation and transcriptional silencing. Indeed recent evidences indicate that Sir2 proteins are involved in the regulation of p53–dependent apoptotic response via deacetylation of p53 proteins. Histone deacetylase inhibition is known to increase the sensitivity of different topoisomerase I inhibitors in different cancer cells . Role of Sir2 is established in PCD of leishmanial cells as well as different kinetoplastids.[46] Further studies are needed to explore the role of Sir2 in topoisomerase mediated cell death. The nuclear topoisomerase I enzyme is responsible for maintaining DNA topology. By breaking and religating DNA strands, they reduce torsional strain during cellular metabolism. Topoisomerase mediated DNA lesions are also regulated by tyrosyl DNA phosphodiesterase in mammalian cells. Tdp 1 was first noted as an enzyme in yeast with activity that specifically cleaves the phosphodiester bond in topoisomerase I—DNA complex.[47] Recently a novel anti apoptotic factor, Bax inhibitor 1 (BI-1) was identified which is located at the endoplasmic reticulum and interacts with Bcl-2 but not Bax in yeast cells. When over expressed in mammalian cells, it suppresses apoptosis induced by Bax, growth factor withdrawal or various drugs but not by Fas. Elucidation of the role of Tdp 1, BI-1 and EF-1α in apoptosis of unicellular kinetoplastid parasites sheds new sight on the cellular complexity of the seemingly simple protozoa.

In the future these kinetoplastid protozoan parasites promise the identification of components of the basic, evolutionarily ancient stages of apoptosis. These organisms offer the opportunity for easy screening of substances acting directly on these basic components without being diverted by a complex upstream network. In addition to providing insights into the origin of PCD, these kinetoplastid parasites also lead to new targets for treatment and prevention of parasitic diseases.

Acknowledgements

This work was supported by the grants from Network Project SMM-003 of Council of Scientific and Industrial Research (CSIR), Government of India to H.K.M. Council of Scientific and Industrial Research (CSIR), Government of India supported N.S. with a Senior Research Fellowship.

References

1. Wang JC. Cellular roles of DNA topoisomerases: a molecular perspective. Nat Rev Mol Cell Biol 2002; 3:430-440.
2. Cortes F, Pastor N, Mateos S et al. Roles of DNA topoisomerases in chromosome segregation and mitosis. Mutat Res 2003; 543:59-66.
3. Stewart L, Ireton GC, Champoux JJ. Reconstitution of human topoisomerase I by fragment complementation. J Mol Biol 1997; 269:355-372.
4. Berger JM. Structure of DNA topoisomerases. Biochim Biophys Acta 1998; 1400:3-18.
5. Das BB, Sen N, Ganguly A et al. Reconstitution and functional characterization of the unusual bi-subunit type I DNA topoisomerase from Leishmania donovani. FEBS Lett 2004; 565:81-88.
6. Villa H, Otero Marcos AR, Reguera RM et al. A novel active DNA topoisomerase I in Leishmania donovani. J Biol Chem 2003; 278:3521-3526.
7. Das BB, Sen N, Dasgupta SB et al. N-terminal region of the large subunit of Leishmania donovani bisubunit topoisomerase I is involved in DNA relaxation and interaction with the smaller subunit. J Biol Chem 2005; 280;16335-16344.

8. Das BB, Sen N, Roy A et al. Differential induction of Leishmania donovani bi-subunit topoisomerase I-DNA cleavage complex by selected flavones and camptothecin: activity of flavones against camptothecin-resistant topoisomerase I. Nucleic Acids Res 2006; 34:1121-1132.

9. Cheng C, Kussie P, Pavletich N et al. Conservation of structure and mechanism between eukaryotic topoisomerase I and site-specific recombinases. Cell 1998; 92:841-850.

10. Pasion SG, Hines JC, Aebersold R et al. Molecular cloning and expression of the gene encoding the kinetoplast-associated type II DNA topoisomerase of Crithidia fasciculata. Mol Biochem Parasitol 1992; 50:57-67.

11. Das A, Dasgupta A, Sharma S et al. Characterisation of the gene encoding type II DNA topoisomerase from Leishmania donovani: a key molecular target in antileishmanial therapy. Nucleic Acids Res 2001; 29:1844-1851.

12. Strauss PR, Wang JC. The TOP2 gene of Trypanosoma brucei: a single-copy gene that shares extensive homology with other TOP2 genes encoding eukaryotic DNA topoisomerase II. Mol Biochem Parasitol 1990; 38:141-150.

13. Fragoso SP, Goldenberg S. Cloning and characterization of the gene encoding Trypanosoma cruzi DNA topoisomerase II. Mol Biochem Parasitol 1992; 55:127-134.

14. Gaziova I, Lukes J. Mitochondrial and nuclear localization of topoisomerase II in the flagellate Bodo saltans (Kinetoplastida), a species with noncatenated kinetoplast DNA. J Biol Chem 2003; 278:10900-10907.

15. Sengupta T, Mukherjee M, Mandal C et al. Functional dissection of the C-terminal domain of type II DNA topoisomerase from the kinetoplastid hemoflagellate Leishmania donovani. Nucleic Acids Res 2003; 31:5305-5316.

16. Das A, Mandal C, Dasgupta A et al. An insight into the active site of a type I DNA topoisomerase from the kinetoplastid protozoan Leishmania donovani. Nucleic Acids Res 2002; 30:794-802.

17. Sengupta T, Mukherjee M, Das R et al. Characterization of the DNA-binding domain and identification of the active site residue in the 'Gyr A' half of Leishmania donovani topoisomerase II. Nucleic Acids Res 2005; 33:2364-2373.

18. Ferguson M, Torri AF, Perez-Morga Ward DC et al. DNA replication: mechanistic differences between Trypanosoma brucei and Crithidia fasciculata. J Cell Biol 1994; 126:631-639.

19. Wang Z, Drew ME, Morris JC et al. Asymmetrical division of the kinetoplast DNA network of the trypanosome. EMBO J 2002; 21:4998-5005.

20. Myler PJ, Glick D, Feagin JE et al. Structural organization of the maxicircle variable region of Trypanosoma brucei: identification of potential replication origins and topoisomerase II binding sites. Nucleic Acids Res 1993; 21:687-694.

21. Sloof P, deHaan A, Eier W et al. The nucleotide sequence of the variable region in Trypanosoma brucei completes the sequence analysis of the maxicircle component of mitochondrial kinetoplast DNA. Mol Biochem Parasitol 1992; 56:289-299.

22. Hajduk Sl, Klein VA, Englund PT. Replication of kinetoplast DNA maxicircles Cell 1984; 36:483-492.

23. Perez-Morga D, Englund PT. The structure of replicating kinetoplast DNA networks. J Cell Biol 1993; 123:1069-1079.

24. Wang JC. DNA topoisomerases as targets of therapeutics: an overview. Adv Pharmacol 1994; 29A:1-19.

25. Ulukan H, Swaan PW. Camptothecins: a review of their chemotherapeutic potential. Drugs 2002; 62:2039-2057.

26. Kim DK, Lee N. Recent advances in topoisomerase I-targeting agents, camptothecin analogues. Mini Rev Med Chem 2002; 2:611-619.

27. Grabowski DR, Holmes KA, Aoyama M et al. Altered drug interaction and regulation of topoisomerase IIbeta: potential mechanisms governing sensitivity of HL-60 cells to amsacrine and etoposide. Mol Pharmacol 1999; 56:1340-1345.

28. Snyder RD. Evidence from studies with intact mammalian cells that merbarone and bis(dioxopiperazine)s are topoisomerase II poisons. Drug Chem Toxicol 2003; 26:15-22.

29. Lewy DS, Gauss CM, Soenen DR et al. Fostriecin: chemistry and biology. Curr Med Chem 2002; 9:2005-2032.

30. Sordet O, Khan QA, Kohn KW et al. Apoptosis induced by topoisomerase inhibitors. Curr Med Chem Anticancer Agents 2003; 3:271-290.

31. Mills JC, Stone NL, Pittman RN. Extranuclear apoptosis. The role of the cytoplasm in the execution phase. J Cell Biol 1999; 146:703-708.

32. Wanderley JL, Benjamin A, Real F et al. Apoptotic mimicry: an altruistic behavior in host/Leishmania interplay. Braz J Med Biol Res 2005; 38:807-812.

33. Debrabant A, Lee N, Bertholet S et al. Programmed cell death in trypanosomatids and other unicellular organisms. Int J Parasitol 2003; 33:257-267.
34. Szallies A, Kubata BK, Duszenko M. A metacaspase of Trypanosoma brucei causes loss of respiration competence and clonal death in the yeast Saccharomyces cerevisiae. FEBS Lett 2002; 517:144-150.
35. Luo J, Nikolaev AY, Imai S et al. Negative control of p53 by Sir2alpha promotes cell survival under stress. Cell 2001; 107:137-148.
36. Duttaroy A, Bourbeau D, Wang XL et al. Apoptosis rate can be accelerated or decelerated by overexpression or reduction of the level of elongation factor-1 alpha. Exp Cell Res 1998; 238:168-176.
37. Solari E, Droin N, Batteiab A et al. Positive and negative regulation of apoptotic pathways by cytotoxic agents in hematological malignancies. Leukemia 2000; 14:1833-1849.
38. Ravagnan L, Roumier T, Kroemer G. Mitochondria, the killer organelles and their weapons. J Cell Physiol 2002; 192:131-137.
39. Nagata S. Apoptosis by death factor. Cell 1997; 88:355-365.
40. Salvesen GS, Dixit VM. Caspase activation: the induced-proximity model. Proc Natl Acad Sci 1999; 96:10964-10967.
41. Sen N, Das BB, Ganguly A et al. Camptothecin induced mitochondrial dysfunction leading to programmed cell death in unicellular hemoflagellate Leishmania donovani. Cell Death Differ 2004; 11:924-936.
42. Sen N, Das BB, Ganguly A et al. Camptothecin-induced imbalance in intracellular cation homeostasis regulates programmed cell death in unicellular hemoflagellate Leishmania donovani. J Biol Chem 2004; 279:52366-52375.
43. Mittra B, Saha A, Chowdhury AR et al. Luteolin, an abundant dietary component is a potent anti-leishmanial agent that acts by inducing topoisomerase II-mediated kinetoplast DNA cleavage leading to apoptosis. Mol Med 2000; 6:527-541.
44. Sen N, Das BB, Ganguly A et al. Intracellular ATP level regulates apoptosis-like death in luteolin induced dyskinetoplastid cells. Exp Parasitol 2006.
45. Singh G, Jayanarayan KG, Dey CS. Novobiocin induces apoptosis-like cell death in topoisomerase II over-expressing arsenite resistant Leishmania donovani. Mol Biochem Parasitol 2005; 141:57-69.
46. Ouaissi A. Apoptosis-like death in trypanosomatids: search for putative pathways and genes involved. Kinetoplastid Biol Dis, 2003:2-5.
47. Yang SW, Burgin AB Jr, Huizenga BN et al. A eukaryotic enzyme that can disjoin dead-end covalent complexes between DNA and type I topoisomerases. Proc Natl Acad Sci 1996; 93:11534-11539.

CHAPTER 6

DNA Metallo-Intercalators with Leishmanicidal Activity

Maribel Navarro,* Gonzalo Visbal and Edgar Marchán

Abstract

In the present chapter we focus our attention on the use of metal complexes for the treatment of leishmaniasis, especially those that have leishmanicidal activity which could be associated with their interaction with the parasitic DNA. Furthermore, we revise current knowledge of leishmaniasis, including PCD-inducing drugs used clinically and those currently in the experimental and evaluation phases.

Introduction

Leishmaniasis comprises a group of diseases with extensive morbidity and mortality in most developing countries. They are caused by species of the genus *Leishmania* (Sarcomastigophora, Kinetoplastida) and range from self-healing cutaneous leishmaniasis (CL) to progressive muco-cutaneous infections (MCL) or fatal disseminating visceral leishmaniasis (VL). While CL poses basically cosmetic problems and MCL leads to painful disfiguration, social stigmatisation and often severe secondary infections, VL is generally lethal if left untreated. According to the World Health Organization (TDR, 2005), leishmaniasis currently affects some 12 million people worldwide and there are 2 million new cases per year with a tendency to increase. Moreover, it is estimated that approximately 350 million people live at risk of infection with *Leishmania* parasites. Leishmaniases are prevalent in 88 countries throughout the world in tropical to Mediterranean climate zones, including 22 in the New World and 66 in the Old World; of these, 72 are developing countries.[1] CL is endemic in Iran, Saudi Arabia, Syria, Afghanistan and in some South American countries. More than 90% of the VL cases worldwide have been registered from India, Bangladesh, Indonesia and the Sudan. In Mediterranean Europe, poor-health communities and certain risk groups such as intravenous drug abusers sharing needles and immunodeficient persons (e.g., AIDS-patients) are severely affected. *Leishmania*/HIV co-infections have increased in Mediterranean countries, where up to 70% of potentially fatal VL cases are associated with HIV infection and up to 9% of AIDS cases suffer from newly acquired or reactivated VL.[2,3]

Protozoa of the genus *Leishmania* are obligate intracellular parasites of mononuclear phagocytes in vertebrate hosts.[4] The pathogen requires two different hosts to complete its biological cycle (Fig. 1): an insect vector (sandflies of the genus *Phlebotomus* in the Old World and *Lutzomyia* in the New World) and a vertebrate host (e.g., humans, rodents, dogs). Within the insect vector, the parasite exists as an extracellular, motile flagellate in the gut. During a blood meal, promastigotes are discharged in the bloodstream and are rapidly phagocytized by Langerhans cells, macrophages, monocytes, or, transiently, by neutrophils. Within these cells they reside in compartments originating

*Corresponding Author: Maribel Navarro—Instituto Venezolano de Investigaciones Científicas (IVIC). Carretera Panamericana, KM 11, Altos de Pipe, Centro de Química, Caracas1020-A, Venezuela. Email: mnavarro@ivic.ve

Programmed Cell Death in Protozoa, edited by José Manuel Pérez Martín.
©2008 Landes Bioscience and Springer Science+Business Media.

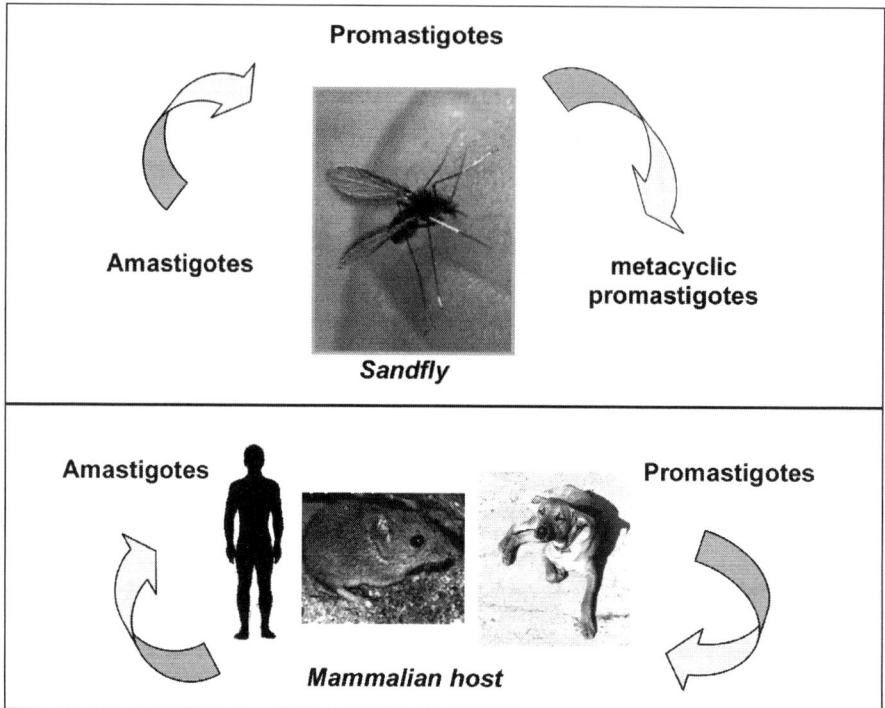

Figure 1. Schematic representation of the life cycle of *Leishmania spp* parasites.

from the plasma membrane, the phagosomes and transform into nonmotile amastigotes. Lysosomes readily fuse with the phagosomes, but *Leishmania* amastigotes not only resist phagolysosomal enzymes, they also thrive and multiply within the acidic, hydrolase-rich parasitophorous vacuole. Massive amastigote multiplication leads to host cell disruption and release of amastigotes to infect newly recruited host cells. Although the parasite is sensitive to humoral defense mechanisms such as antibodies or the complement system, its intracellular habitat offers almost complete protection. Only if the macrophage is activated, the parasites may be killed and then degraded by the host cell. Ingestion of infected peripheral monocytes during the blood meal by a female sandfly completes the biological cycle.[5]

Programmed Cell Death (PCD) in Trypanosomatids

To survive successfully and multiply within two disparate biological environments (digenic life cycle), *Leishmania* parasites must undergo profound biochemical and morphological adaptations.

The ability of this parasite to undergo differentiation from the promastigote to the amastigote form is crucial for its pathogenesis. Furthermore, during its life cycle, *Leishmania* grows in diverse and hostile environmental conditions and has to control it own growth as well as preserve individuals that are fit to continue the infective cycle.[6]

PCD is an essential part of cell biology and is thought to have evolved not only to regulate growth and development in multicellular organisms but also to guard against viral infections and the emergence of cancer. However, recent studies, which showed the existence of PCD in unicellular organisms, have postulated a functional role of PCD in the biology of unicellular organisms. Furthermore, in the light of a reported nonapoptotic programmed cell death, parapoptosis, it is important to understand the type of PCD that exists in unicellular organisms such as Trypanosomatids.[7]

It has been postulated that in order to promote and maintain clonality within the population, the *Trypanosomatids* must have developed an altruistic mechanism to control growth.[8] PCD has been shown to be involved in the control of cell proliferation of *T. cruzi* in vitro[9] and in the insect vector midgut.[10] Features of PCD also have been observed in procyclic insect form in *T. brucei rhodesiense* upon treatment in vitro with lectin.[11] Ca^{2+} modulation by heat shock during differentiation in *L. amazonensis* also induced PCD.[12] A cell death process akin to PCD was also demonstrated to exist in slime mold *Dictyostelium discoideum*, an organism that exists both in a unicellular and multicellular form, thus reinforcing the idea that the pathway for PCD must have evolved before the evolution of multicellularity.[6,13] All these observations point towards the existence of a PCD pathway in *Trypanosomatids*.

It has been demonstrated some features characterizing PCD in the unicellular protozoan parasite *Leishmania donovani*, the causative agent of visceral Leishmaniasis. They reported that PCD is initiated in the stationary phase cultures of promastigotes and in both actively growing cultures of axenic amastigotes and promastigotes upon treatment with anti Leishmanial drugs (Pentostam and amphotericin B). The features of PCD in *L. donovani* promastigotes were nuclear condensation, nicked DNA in the nucleus, DNA ladder formation, increase in plasma membrane permeability, a decrease in the mitochondrial membrane potential ($\Delta\Psi_m$) and induction of a PhiPhiLux (PPL)-cleavage activity. PCD in both stationary phase culture and upon induction by amphotericin B resulted first in the decrease of mitochondrial membrane potential followed by simultaneous change in plasma membrane permeability and induction of PPL-cleavage activity, demonstrating that the characteristic features of PCD exist in unicellular protozoan *Leishmania donovani*.[7]

Elucidation of the molecular events linked to the apoptotic death of *Leishmania spp.* might help define a more comprehensive view of the cell death machinery in terms of evolutionary origin as well as allowing the identification of new target molecules for chemotherapeutic drug development and therapeutic intervention.[14]

PCD by Antileishmanial Drugs

Treatment available for leishmaniasis is far from ideal. The classic first line of treatment is based on pentavalent antimonials (Sb(V)) with sodium stibogluconate (Pentostam*) recommended for intravenous and intramuscular administration and meglumine antimoniate (Glucantime) for intramuscular administration. Both drugs exist only in their parental forms and have been used for over 60 years; their history and use[1,15] as well as the pharmacology of leishmaniasis[1] have been excellently reviewed. Pentavalent antimony is generally regarded as a prodrug requiring conversion to a trivalent form in order to be cytotoxic and this reduction takes place in the parasite or the host cells or both. Antimonials are thought to act by inhibiting metabolic pathways although the mode of action is still poorly understood.[14]

Although these antimonials are successfully employed worldwide, they have been shown to have severe side effects such as cardiotoxicity, reversible renal insufficiency, pancreatitis, anemia, leukopenia, the appearance of rashes, headaches, abdominal pains, nausea, vomiting, thrombocytopenia and transaminase elevation. Furthermore, the dramatic increase in the number of cases that do not respond to these drugs represents a critical limitation for the treatment of these parasitic diseases.[16]

There are several reports showing that *Leishmania* PCD occurs in response to pentavalent antimonials. Most of these studies were carried out in vitro using axenic amastigotes generated in the laboratory by subjecting the promastigotes to changes in pH and temperature. Although not totally resembling intracellular amastigotes, these organisms are frequently used as drug testing models.[17]

It has been found that during the late log growth phase of *L. donovani* axenic amastigotes, treatment with increasing concentrations of Sb(V) produced a significant induction of caspase-like activity. On the contrary, no significant caspase-like activity was observed in promastigotes upon treatment with Sb(V) even at a higher concentration or when treated for a longer period of time, showing that there is a difference in how the different forms of the parasite respond to Sb (V). In contrast to the above observations, no caspase-like activity upon treatment of axenic amastigotes of *L. infantum* with trivalent antimonial compound Sb(III) was detected although the presence of apoptotic features such as DNA fragmentation was seen.[7,18] Therefore, these observations suggest the possibility of the existence of a caspase independent apoptotic pathway in *Leishmania spp* similar to that described from multicellular organisms[19,20] where the release of death factors from mitochondria has been shown to cause apoptosis without involving caspases.

Based on the capacity of *L. donovani* to survive within the host cell parasitophorous vacuoles as nonmotile amastigotes determines disease pathogenesis, thus apoptosis in the intracellular forms is very important. It has been found that potassium antimony tartrate, the active form of Sb (V), kills intracellular *L. donovani* amastigotes by PCD characterized by nuclear DNA fragmentation and the externalization of phosphatidylserine.[21]

When Sb(V) fails, Amphotericin B is a recommended second line treatment for visceral, cutaneous and mucocutaneous leishmaniasis.[22-26] Amphotericin B, a pore forming polyene antibiotic, induces caspase-like activity in both axenic amastigotes and promastigotes of *L. donovani* resulting in DNA fragmentation within 2 h in vitro,[7] suggesting that the PCD effect of amphotericin B is faster compared to antimony. This drug is administered over a period of 4-6 hours by slow intravenous infusion, starting at a dose of 0.1 mg/kg and gradually increasing to 1 mg/kg, under strictly regulated conditions.

The side effects include fever, nausea, vomiting malaise, anemia, phlebitis, hypokalemia, hypomagnesemia and principally nephrotoxicity. Several amphothericin B lipid formulations with much lower toxicities but at a higher cost than the free drugs have been developed and have proved to be useful in the treatment of visceral leishmaniasis.[25,26] The treatment period is short, reducing hospitalization costs and are used as the drugs of preference in developed countries. In developing countries these high cost formulations remain inaccessible for routine use.[27]

Another drug, the aromatic diamidine pentamidine, initially developed for use against *Trypanosoma brucei*, is considered an alternative to pentavalent antimony for the treatment of leishmaniasis. Concurrent inhibition of respiratory chain complex II with pentamidine administration increases the cytotoxicity of the drug leading to death by PCD.[28]

The pentamidine isethionate has had some success in the treatment of visceral leishmaniasis. Unfortunately the cure rate declined between the early 1980s and the 1990s due to the resistance of *Leishmania* to pentamidine.[1,29] Cumulative effects, which often limit dose or frequency of administration, include weakness, nausea, vomiting and abdominal pain, which may indicate pancreatic damage. The unusually high rate of hyperglycemia (50%) associated with its use has been attributed to the high rate of pancreatic fibrosis. Others have also suggested its relation to hypotention, tachycardia and cardiotoxicity.[30]

Recently, in the search for oral drugs which eliminate complications associated with parenteral administration, Miltefosine was discovered, a promising new drug designed for cancer treatment belonging to the alkilphosphocholines group and currently representing the first oral treatment for visceral leishmaniasis. It induces PCD-like death in *L. donovani* showing nuclear DNA condensation and DNA fragmentation with accompanying ladder formation.[31] In experimental and clinical assays it was found to have an effectivity of 94-97%. Nevertheless, the drug produced severe gastrointestinal side effects and can not be administered during gestation due to its teratogenic potential.[32,33] The high cost of this drug has been compensated by the development of strategic alliances between the Government of India, Zentaris and WHO/TDR to be used as the first line of treatment for leishmaniasis in the Indian subcontinent.[34]

Figure 2 shows the clinical antileishmanial drugs with the ability to induce PCD above mentioned.

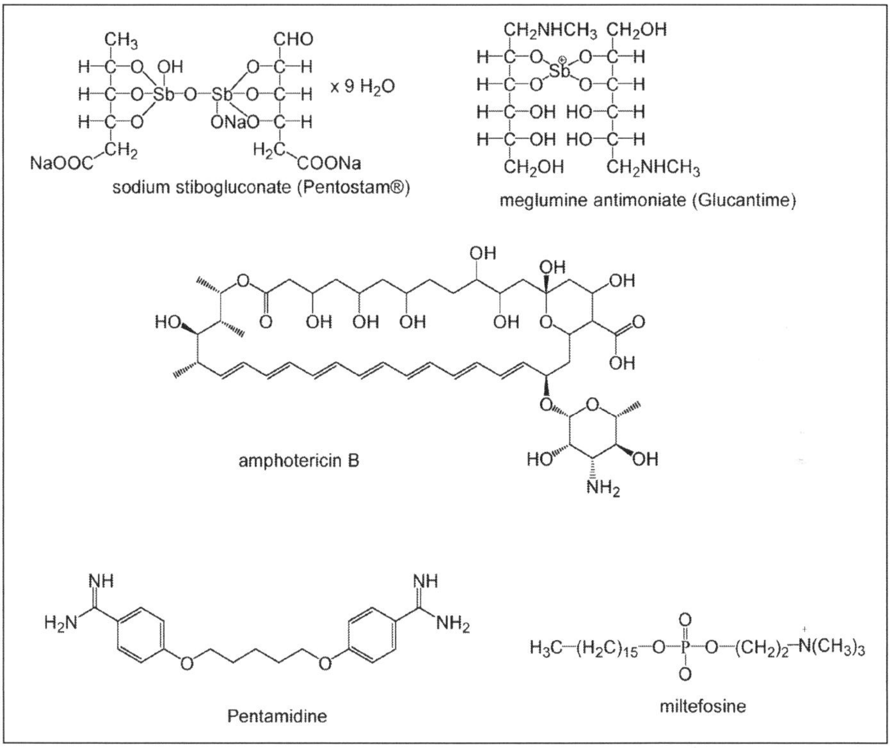

Figure 2. Clinical antileishmanial drugs with the ability to induce PCD.

Other Leishmanicidal Drugs

Other drugs, allopurinol, rifampicin, dapsone, chloroquine and nifurtimox, have found favor in some studies, although they have not been widely accepted for clinical use.[35-44]

Currently, there is a renewed interest in the investigation of plant products with antileishmanial activities, due at least in part, to the identification of licochalcone A as a potential drug against *Leishmania, Trypanosoma* and *Plasmodium* parasites.[45] Recently, a study of a new model for the administration of transchalcone biodegradable polymers (polylactic acid [PLA] and polylactic/glycolic acid [PLGA]) for the treatment of experimental leishmaniasis was published that supports this idea.[46] Advances in the research of natural products for the treatment of leishmaniasis have been recently reviewed[16,47-50] and provide a valuable source of new medicinal agents. To date, only berberine has been used clinically for the treatment of cutaneous leishmaniasis, whereas chimamine D and 2-n-propylquinoline have reached the clinical evaluation phase for the treatment of cutaneous leishmaniasis.[51-54] Both of these natural products have also proved to be active against visceral leishmaniasis in mice when administered orally.

Another interesting family of compounds with efficacy against trypanosomatids are the sterol biosynthesis inhibitors (SBIs), which can be divided into three groups: (1) the allilamines, of which terbinafine is a potent inhibitor of squalene-2,3-epoxidase[55] and is an effective agent against *Leishmania sp.*[56-59] (2) Imidazole (ketoconazole, clotrimazole) and triazoles (fluconazole, itraconazole) have proved effective against several *Leishmania sp.*[60] As in fungi, they exert their effect by blocking the cytochrome P450 dependent C14α-demethylase in promastigotes and amastigotes with a resulting decline in the normal complement of endogenous C_{28} and C_{29} sterols and an accumulation of various 14α-methyl sterols altering the normal permeability and fluidity of the parasite membrane, with secundary consequences for membrane-bound enzymes.[1,58-66] Ketoconazole and itraconazole are the only compounds within this group that have been used systemically for cutaneous leishmanaisis treatment. Clinical studies have demonstrated beneficial effects for patients suffering cutaneous leishmanaisis when ketoconazole is administered orally at a dose 200-400 mg/day varying from 8-16 weeks. Itraconazole has also been used at a dose of 200 mg/day for 4-8 weeks, but data are scare and controversial. The adverse effects that these azoles possess are nausea, abdominal pains, headaches, fever, usually mild reversible hepatotoxicity, lowering of testosterone values (70%), but without a reduction in libido or beard growth. (3) the azasterols (with a nitrogen located in the side chain) that block the C-24 alkylation reaction in sterol biosyntesis[55,67] have been tested against trypanosomatids.[67-70] Of these, the 20-piperdin-2-yl--α-pregnane-3β,20R-diol (22,26 azsterols) halted promastigote growth in the parasite *Leishmania donovani* at concentrations as low as 12 nM while being lethal at concentrations higher than 10 μM. The inhibition of the C24-methylation reaction caused a depletion of the C_{28}-sterols and the accumulation of C_{27}-sterols, mainly cholesta-5,7,24-trien-3β-ol and some cholesta-7,24-dien-3β-ol, together with cholesterol derived from the culture medium.[70] Recently, studies of the synthesis of new azasterols and their evaluation against *Leishmania spp.* showed activity at micromolar and nanomolar concentrations.[69,70] The selected organic compounds before mentioned are shown in Figure 3.

The above mentioned classic and new experimental drugs for the treatment of leishmaniasis have several drawbacks, such as the need for long periods of medication, renal disruption or other side effects, the emergence of resistance and high costs (lipid formulations of amphotericin B). Thus, the search for new and more effective chemotherapeutic agents against leishmaniasis with fewer or no side effects, continues. To this end, the rational design of new experimental antileishmanial drugs is an important goal.

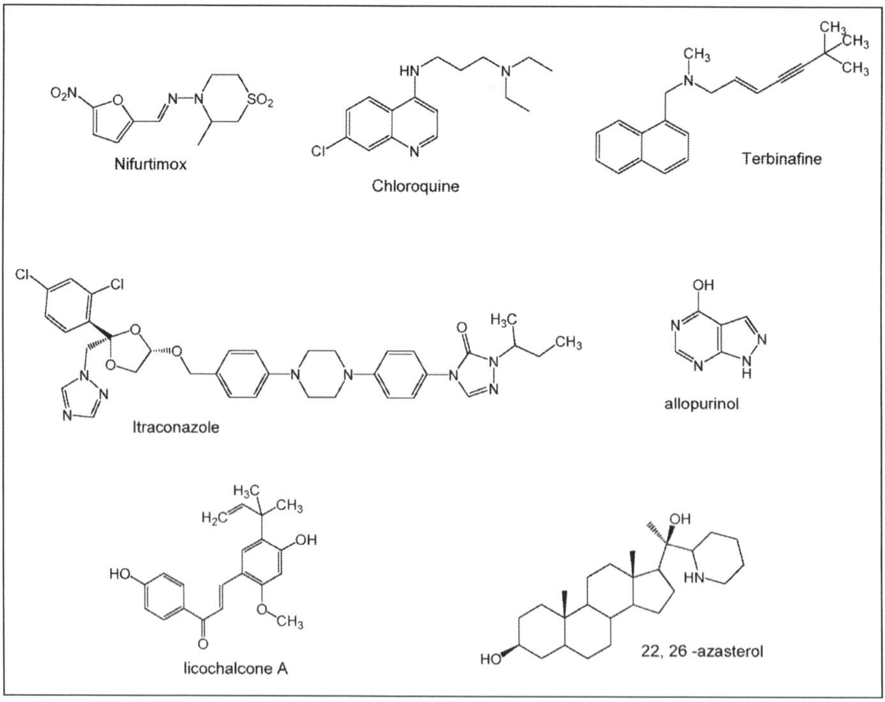

Figure 3. Selected organic compounds tested against *Leishmania spp* parasites.

Transition Metal Complexes as Antileishmanial Agents

Transition metal-based chemotherapeutic drugs have attracted considerable attention in the last several years, mainly as a result of the success of cis-platin and the new generation of Pt complexes as well as intense interest and research to investigate other metals such as titanium and ruthenium as antitumoral agents.[71-79] However, despite this research progress in the inorganic medicinal chemistry field to fight against cancer, the use of metal containing drugs as antiparasitic agents have been scarcely explored. Indeed, this approach seems very attractive due to there is (1) an urgent need for new effective and nontoxic antiparasitic drugs, and (2) an immense number of rational combinations among appropriate organic ligands and different transition metals. This approach makes use of the concept of metal-drug synergism, which is the enhancement of the activity of the parental organic drug due to binding to the metal ion; this is possibly related to the stabilization of the drug, which leads to a longer time of residence in the organism, allowing it to reach the biological targets more efficiently and may also result in a decrease in the toxicity of the metal ion caused by its complexation with organic drugs, which makes the metal less available for undesirable reactions that lead to toxicity.[80] Although the design of metal complexes with good therapeutic indexes is still very empirical, the metal-drug synergism is a powerful tool which could provide effective treatment for leishmaniasis in the future.

The use of transition metal complexes as potent chemotherapeutic agents against leishmaniasis remains a priority since the drugs currently available are not totally safe and active. Moreover, the appearance of drug-resistant strains of *Leishmania spp.* justifies the screening of new compounds. Organometallic compounds of pentamidine with the general formula $[M_2(L_2)(\text{pentamidine})]$,[+2] $M = Ir$, Rh; $L_2 = 1,5$-cyclooctadiene COD, 1,3-1,5-cyclooctatetraene (COT) or $(CO)_2$, with several counteranions such as nitrate (Fig. 4A) have been shown to be active against *L. donovani* promastigotes, some of these being even more active than pentamidine isethionate. The related compound $[Ir_2(COT)_2(\text{pentamidine})][\text{alizarin red}]_2$ (Fig. 4B) was also shown to be at least twice as active as pentamidine isethionate against the amastigote form of *L. donovani*[81-83] and a synergistic effect was noted when this complex was administered in combination with pentamidine, amphotericin B or paromomycin.[84]

Pt(II)-pentamidine complexes appear to be less active than Rh(I) and Ir(I) analogs against amastigotes of *L. donovani*.[85] In a previous publication this group of researchers have also reported the activity against *Leishmania donovani* of several platinum, osmium and rhodium complexes such as $[cis\text{-}Pt(II)(1,2\text{-diaminocyclohexane})(2,5\text{-dihydroxybenzenesulfonate})_2]$ $[RhCl(CO)_2(2\text{-aminobenzothiazole})]$ and $[RhCl(CO)_2(2\text{-methylbenzothiazole})]$ and Os(III) complexes with nitroimidazole dithiocarbamates, benznidazole dithiocarbamates and related ligands which are displayed a moderate activity against *L. donovani*.[86-87] Simple salts such as zinc sulphate have been tested clinically for treatment of cutaneous leishmaniasis with very promising cure rates of 96.9%, using oral doses of 10 mg/kg in 45 days of treatment.[88]

The metallointercalator of DNA (2,2':6'2"-terpyridine)platinum (II) (Fig. 4C) and analogs have shown remarkable antileishmanial activity, the most effective compound caused 100% growth inhibition of the intracellular amastigote forms of *L. donovani* at a concentration of 1 μM.[89] These complexes exploit the intercalative DNA properties of the terpyridine ligand along with the covalent binding ability of the Pt(II) center. The highest activity against *L. donovani* was found for the compound with *p*-bromophenyl substituents in the 4'-terpyridine position and NH_3 as the ancillary hydrolysable ligand.

Water soluble cationic trans-platinum complexes are active against *L. infantum*. The four trans-Pt complexes reported exhibited an antileishmanial activity higher than that of trans-platin. Moreover, two of these complexes (Fig. 4 D,E) showed leishmanicidal activity between 1.6 and 2.5- times higher than cisplatin. The cytometry data indicated that the higher leishmanicidal activity of these two Pt-complexes relative to cisplatin is related to their greater ability to induce both PCD and blocking of cell division in G2/M. It may be concluded that both trans Pt-complexes bind to nuclear DNA in a way different to that of cis-platin.[90]

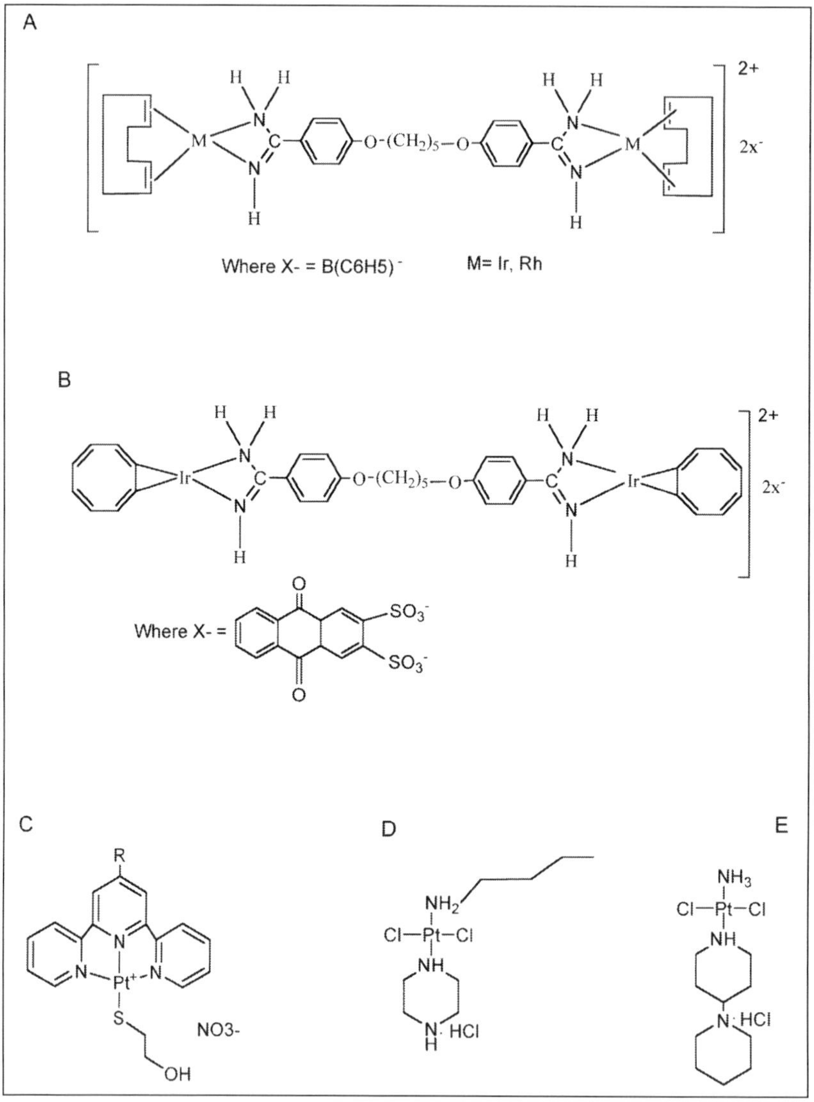

Figure 4. Selected metal complexes tested against *Leishmania spp* parasites.

DNA as Leishmanicidal Target

Over the past two decades there has been substantial interest in the study of DNA metal-lointercalators due to their possible uses as new therapeutic agents. Here we review the design and development of metallointercalators that bind and react with DNA. Intercalators are small molecules that contain a planar aromatic heterocyclic functionality which can insert and stack between the base pairs of double helical DNA[91] without forming covalent binding. The first met-allointercalators established was the square planar platinum (II) complex containing an aromatic heterocyclic ligand that could bind to DNA by intercalation.[92] The only recognized forces that maintain the stability of the DNA-intercalators complex, even more than DNA alone are van der Waals, hydrogen bonding, hydrophobic and or change transfer forces.[93-94]

Many of the physical properties of DNA are changed due of interaction with the metallointercalators, the alteration of their structure allows the application of a wide range of techniques for its measurement, such as ultraviolet spectroscopy which is a useful and important techniques for obtaining the helix-to-coil transition temperature (T_m) also called the melting or denaturation temperature. This is one of the simplest measurements and is based on the difference in stabilities between the DNA alone and the DNA with intercalator complexes; the stability of DNA when heated is frequently used to measured DNA intercalation. The melting temperature is taken as the temperature at which half of the DNA has denatured, which is pH dependent. Hydrodynamic properties such as viscosity[95] and ethidium bromide displacement[96] are more selective methods for the determination of DNA-intercalation. Circular dichroism (CD) and fluorescence spectroscopy are also frequently used methods for measuring this interaction.[97-98] More complicated, but also useful, are mass detection using an electrospray and nuclear magnetic resonance (NMR) technique. Finally, the technique that provides the most unequivocal results is X-ray diffraction.[99] However, these last two are among the most consuming and expensive of the available methods.

The studies of DNA metallointercalators as possible chemotherapeutic agents have focused principally on the search for complexes with activity against tumors resistant to cis-platin. it has been proposed that intercalating ligands could act as carriers, increasing the interaction of the complexes with DNA by minimizing the exposure of the metal to inactivating cellular nucleophiles such as thiols.[100]

Since metabolic pathways of kinetoplastid parasites are similar to those present in tumor cells,[101] our group has been interested in the design of metallointercalators that could show activity against some of these pathogenic parasites through their interaction with DNA.

Based on all this, we have designed our metals complexes to be DNA intercalator molecules. In order to develop leishmanicidal agents, we used metals of known clinical application such as copper,[102-103] silver[104] and gold[105] with ligands typically employed in the synthesis of metallointercalators such as the planar organic compounds dppz (dipyrido[3,2-a:2',3'-c]phenazine) and dpq (dipyrido[3,2-a:2',3'-h]quinoxoline).

The structures of the copper, silver and gold complexes we have investigated are shown in Figure 5: [Cu(dpq)(NO$_3$)]NO$_3$ (1) and [Cu(dpq)$_2$(NO$_3$)]NO$_3$ (2), [Cu(dppz)(NO$_3$)]NO$_3$ (3), [Cu(dppz)$_2$(NO$_3$)]NO$_3$ (4), [Ag(dpq)$_2$]NO$_3$ (5), [Ag(dppz)$_2$]NO$_3$ (6), [Cu(dppz)$_2$]BF$_4$(7) and [Au(dppz)$_2$]Cl$_3$ (8). These complexes were fully characterized. DNA interaction studies were followed by spectroscopic titrations which allowed us to observe the electronic pertubation produced by the interaction of the metal complexes with DNA finding of hypochromism, bathochromism and their binding constants for these metal complexes with the DNA in the order of 10^3-10^5 M^{-1}. These important parameters have been previously taken as evidence of intercalation, but such data alone are insufficient to rule out alternative mechanisms.[106] Thus we also carried out hydrodynamic measurements such as viscosity, which showed that an increase in viscosity is directly proportional to an increases in the [complex]/[DNA] ratio, in a similar way to that reported for the classical intercalator ethidium bromide.[107] Another experiment was an electrophoresis assay where the complex was mixed with plasmid DNA pUC119 at different concentrations in the dark for 18 h. The increase in the concentration of the complexes (1-8) caused changes in the mobility of the plasmid and the band corresponding to the circular form was observed. These results were similar to those reported for other metallointercalators.[108]

The biological activity of these complexes (1-8) against the promastigote form of *Leishmania (V) braziliensis* (causative agent of the mucocutaneous mode of the disease) and *L. (L) mexicana* (causative agent of the cutaneous mode of the disease), was also evaluated with all of them showing leishmanicidal activity. This biological activity was higher for the metal-dppz complexes than for the metal-dpq complexes. Additionally the complexes with two coordinated molecules of the ligand were more active than those with one. Copper complexes with typical DNA-intercalating ligands, have shown very strong in vitro activity against these two *Leishmania* species *and* it has been demonstrated that their action could be related to their ability to interact with DNA through intercalation.

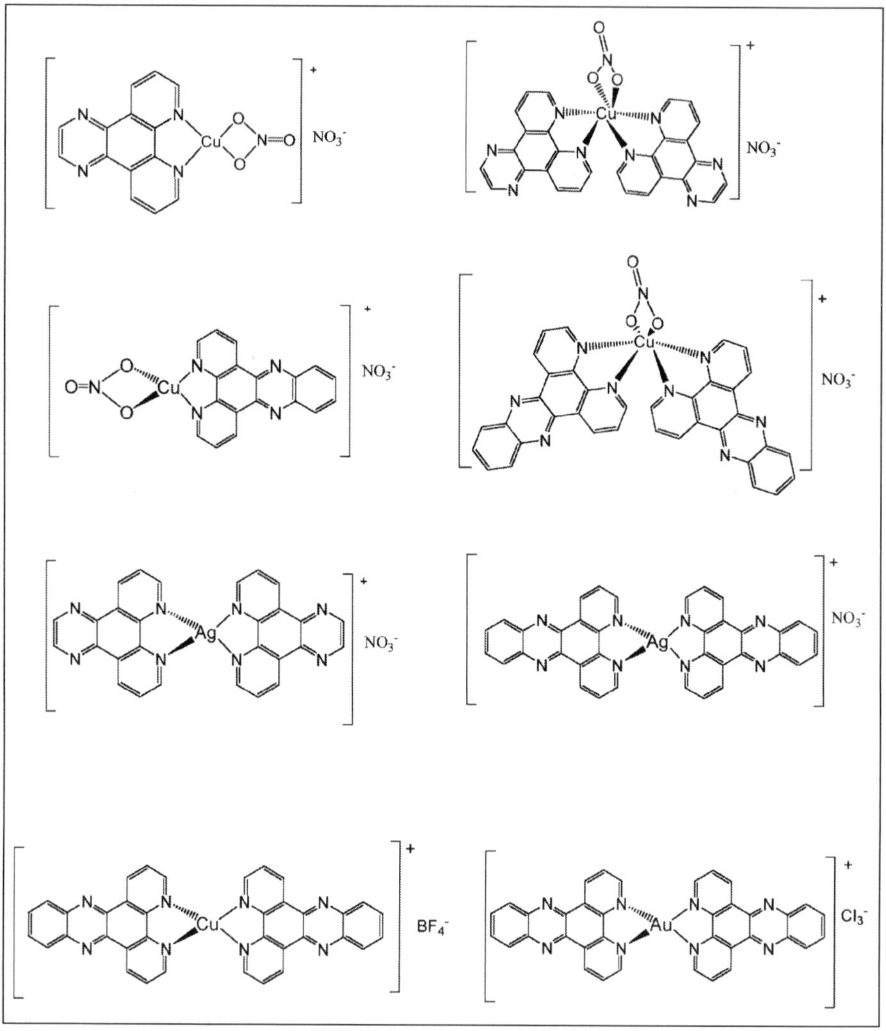

Figure 5. Metallointercalators active against *Leishmania spp.*

[Cu(dppz)$_2$](NO$_3$)$_2$ was the most effective complex in the series of copper (II) complexes, for which the leishmanicidal activity order at 10 µM final concentration at 48 h was: [Cu(dppz)$_2$](NO$_3$)$_2$ > [Cu(dppz)(NO$_3$)](NO$_3$) > [Cu(dpq)$_2$](NO$_3$)$_2$ > [Cu(dpq)(NO$_3$)](NO$_3$). In contrast, [Cu(dppz)$_2$]BF$_4$ which corresponds to Cu(I) complex displayed a higher leishmanicidal effect (LD$_{30}$) at a concentration of only 41 nM after 48 h of treatment, that is, 3 orders of magnitude lower than that observed with Cu (II). Preliminary ultrastructural studies by Transmission Electron Microscopy (TEM) carried out with sublethal concentration (IC$_7$ = 4.1 nM for 48 h) of [Cu(dppz)$_2$]BF$_4$ showed a binucleated cell with disorganized chromatin associated with cytoplasm disorganization and extensive vacuolization (Fig. 6).

On the other hand and also with a 10µM final concentration at 48 h, [Ag(dppz)$_2$]NO$_3$ (LD$_{23}$) and [Ag(dpq)$_2$]NO$_3$ (IC$_{55}$) were less active than [Cu(dppz)$_2$]BF$_4$ and coincided with similar results reported for [Cu(dpq)(NO$_3$)](NO$_3$), demonstrating that the synergism introduced by Cu (I) coordinated to these planar ligands is higher than that observed with Ag.

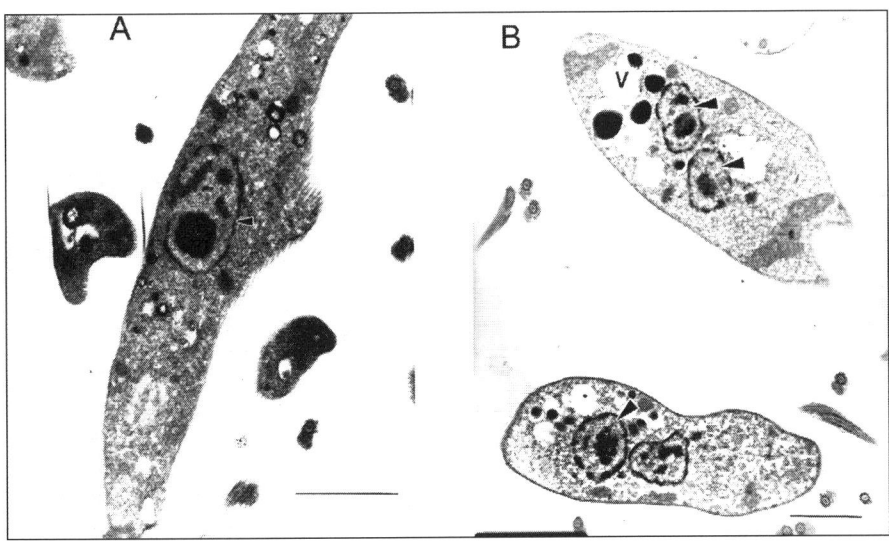

Figure 6. Transmission electron microscope micrograph of *Leishmania mexicana* promastigotes: A) control. B) Parasite treated with [Cu(dppz)2]BF4, 4.1 nM. In (B) arrow heads: nucleus. V. vacuoles. Scale bar 1 mm. (Reprinted from: Mbongo N et al. Acta Tropica 1998; 70:239-245; ©1998 with permission from Elsevier.[84])

Finally, the biological activity when gold was used as the transition metal (corresponding complex, $[Au(dppz)_2]Cl_3$) was the highest of all the complexes evaluated and induced a dose dependent antiproliferative effect with minimal inhibitory concentration (MIC) of 3.4 nM and lethal doses LD_{26} of 17 nM at 48 h.[105] Preliminary ultrastructural studies using TEM carried out with treated parasites at a sublethal concentration ($IC_7 = 0.34$ nM for 24 h) showed polynucleated cells with DNA fragmentation and drastic disorganization of the mitochondria (Fig. 7) (data not published).

In summary, the very potent leishmanicidal activity of these metal complexes could be associated to the PCD cellular processes involving parasite DNA and constitutes a good starting point in the search for new rational and effective chemotherapeutic alternatives for future studies in experimental leishmaniasis models.

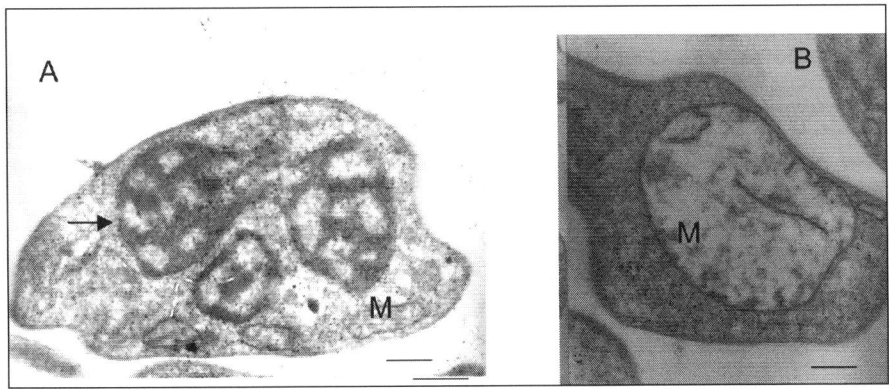

Figure 7. TEM of *L (L) mexicana* promastigotes treated with $Au(dppz)_2Cl_3$ ($IC_7 = 0.34$ nM for 48 h). A) → Nucleus. M, mitochondria. Scale bar 0.7 μm. B) M, mitochondria. Scale bar 0.4 μm.

Acknowledgements

The authors appreciate the support of IVIC and FONACIT grants PEM-2001001621 and G-2005000827. We also wish to thank Dr. C. Hernandez, F. Osborn and Prof. G. Fraile for their critical revision of the manuscript .

References

1. Blum J, Desjeux P, Schwartz E et al. Treatment of cutaneous leishmaniasis among travellers. J Antimicrob Chemother 2004; 53:158-166.
2. Alvar J, Canavata C, Gutiérrez-Solar B et al. Leishmania and human immunodeficiency virus coinfection: the first 10 years. Clin Microbiol Rev 1997; 10:298-318.
3. Desjeux P, Alvar J. Leishmania/HIV co-infections: epidemiology in Europe. Ann Trop Med Parasitol 2003; 97(1):3-15.
4. Alexander J, Russell DG. The interaction of Leishmania species with macrophages. Adv Parasitol 1992; 31:175-254.
5. Alexander J, Satoskar AR, Russell DG. Leishmania species: models of intracellular parasitism. J Cell Sci 1999; 112:2993-3002.
6. Nguewa P, Fuertes M, Valladares B et al. Programmed cell death in trypanosomatids: a way to maximize their biological fitness. Trends Parasitol 2004; 20(8):375-380.
7. Lee N, Bertholet S, Debrabant A et al. Programmed cell death in the unicellular protozoan parasite Leishmania. Cell Death Differ 2002; 9(1):53-64.
8. Welburn SC, Barcinski MA, Williams GT. Programmed cell death in trypanosomatids. Parasitol Today 1997; 13:22-26.
9. Ameisen JC, Idziorek T, Billaut-Mulot O et al. Apoptosis in a unicellular eukaryote (Trypanosoma cruzi): implications for the evolutionary origin and role of programmed cell death in the control of cell proliferation, differentiation and survival. Cell Death Differ 1995; 2:285-300.
10. Welburn SC, Maudlin I, Ellis DS. Rate of Trypanosome killing by lectins in midguts of different species and strains of Glossina. Med Vet Entomol 1989; 3:77-82.
11. Welburn SC, Dale C, Ellis D et al. Apoptosis in procyclic T. B. rhodesiense in vitro. Cell Death Differ 1996; 3:229-236.
12. Moreira ME, Del Portillo HA, Milder RV et al. Heat shock induction of apoptosis in promastigotes of the unicellular organism Leishmania (Leishmania) amazonensis. J Cell Physiol 1996; 167:305-313.
13. Cornillon S, Foa C, Davoust J et al. Programmed cell death in Dictyostelium. J Cell Sci 1994; 107:2691-2704.
14. Shaha C. Apoptosis in Leishmania species & its relevance to disease Pathogenesis. Indian J Med Res 2006; 123:233-244.
15. Berman JD. Chemotherapy for leishmaniasis: biochemical mechanisms, clinical efficacy and future strategies. Rev infect Dis 1988; 10:560-586.
16. Kayser O, Kiderlen AF, Croft SL. In: Atta-ur-Rahman, ed. Studies in Natural Products Chemistry, Bioactive Natural Products (Part G) vol. 26. Amsterdam: Elsevier, 2002:779-848.
17. Sereno D, Cavaleyra M, Zemzoumi K et al. Axenically grown amastigotes of Leishmania infantum used as an in vitro model to investigate the pentavalent antimony mode of action. Antimicrob Agents Chemother 1998; 42:3097-102.
18. Sereno D, Holzmuller P, Mangot I et al. Antimonial-mediated DNA fragmentation in Leishmania infantum amastigotes. Antimicrob Agents Chemother 2001; 45:2064-2069.
19. Matsuyama S, Llopis J, Deveraux QL et al. Changes in intramitochondrial and cytosolic pH: early events that modulate caspase activation during apoptosis. Nat Cell Biol 2000; 2:318-325.
20. Sperandio S, de Belle I, Bredesen DE. An alternative, nonapoptotic form of programmed cell death. Proc Natl Acad Sci USA 2000; 97:14376-81.
21. Sudhandiran G, Shaha C. Antimonial induced increase in intracellular Ca2+ through nonselective cation channels in the host and the parasite is responsible for apoptosis of intracellular Leishmania donovani amastigotes. J Biol Chem 2003; 278:25120-32.
22. Bryceson ADM. Leishmaniasis. In: Wyngaarden JB, Smith Jr LH, eds. Cecil Textbook of Medicine. Edinburgh: Churchill Living-Stone, 1980:815-818.
23. Maegrath B, ed. Clinical Tropical Disease, 7th ed. Vol 12. London: Blackwell Scientific Publication 1980:184.
24. Ganor S. The treatment of leishmaniasis residual with local injections of amphotericin B. Dermatol Int 1967; 6:141-143.
25. Yardley V, Croft SL. A comparison of the activities of three amphotericin B lipidformulations against experimental visceral and cutaneous leishmaniasis. Int J Antimicrob Agents. 2000; 13(4):243-248.

26. Meyerhoff US. Food and Drug Administration approval of AmBisome (liposomal amphotericin B) for treatment of visceral leishmaniasis. Clin infect Dis 1999; 28(1):49-51.

27. Barrat G, Legrand P. Comparison of the efficacy and pharmacology of formulations of ampotericin B used in the treatment of leishmaniasis. Curr Opin Infect Dis 2005; 18(6):527-530.

28. Khalil EA, El Hassan AM, Zijlstra E et al. Treatment of visceral leishmaniasis with sodium stibogluconate in Sudan: management of those who do not respond. Ann Trop Med Parasitol 1998; 92:151-158.

29. Hentzer B, Kobayasi T. The ultra estructural changes of Leishmania tropica after treatment with pent-amidine. Ann Trop Med Parasitol 1977; 71:157-66.

30. Hellier I, Dereure O, Tournillac I et al. Treatment of old world cutaneous leishmaniasis by pentamidine isethionate: An open study of 11 patients. Dermatol 2000; 200(2):120-123.

31. Verma NK, Dey CS. Possible mechanism of miltefosine mediated death of Leishmania donovani. Antimicrob Agents Chemother 2004; 48:3010-3015.

32. Davies C, Kaye P, Croft S et al. Leishmaniasis: new approaches to disease control. British Med J 2003; 326:377-382.

33. Singh S, Sivakumar R. Challenges and new discoveries in the treatment of leishmaniasis. J Infect Chemother 2004; 10(6):307-315.

34. TDR. Tropical Disease Research Progress 2003-2004. Seventeenth Program Report of the UNIFEF/UNDP/WB/WHO Special Program for Research & Training in Tropical Disease, 2005.

35. Soto J, Toledo J, Gutierrez P et al. Treatment of american cutaneous leishmaniasis with miltefosine, an oral agent. Clin Infect Dis 2001; 33:57-61.

36. Saenz RE, Paz HM, Jonhson CM et al. Treatment of american cutaneous leishmaniasis with orally administered allopurinol riboside. J Infect Dis 1989; 160:153-158.

37. Guderian RH, Chico ME, Rogers MD et al. Placebo controlled treatment of Ecuadorian cutaneous leishmaniasis. Am J Trop Med Hyg 1991; 45:92-97.

38. Esfandiarpour I, Alavi A. Evaluating the efficacy of allopurinol and meglumine antimonmiate (Glucantime) in the treatment of cutaneous leishmaniasis. Int J Dermatol. 2002; 41:521-524.

39. D'Oliveira JA, Machado PR, Carvalho EM. Evaluating the efficacy of allopurinol for the treatment of cutaneous leishmaniasis. Int J Dermatol 1997; 36:938-940.

40. Martinez S, Marr JJ. Allopurinol in the treatment of american cutaneous leishmaniasis. New Ingl J Med 1992; 326:741-744.

41. Martinez S, Gonzalez M, Vernaza ME. Treatment of cutaneous leishmaniasis with allopurinol and stibogluconate. Clin Infect Dis 1997; 24:165-169.

42. Kochar DK, Aseri S, Sharma BV et al. The role of rifampicin in the management of cutaneous leishmanaisis. Quart J Med 2000; 93:733-737.

43. Dogra J. A double-blind study on the efficacy of oral dapsone in cutaneous leishmaniasis. Trans Royal Soc Trop Med Hyg 1991; 85:212-213.

44. Sharquie KE, Najim RA, Farjou IB et al. Oral zinc sulphate in the treatment of acute cutaneous leishmaniasis. Clin Exp Dermatol 2001; 26:21-26.

45. Chen M, Christensen SD, Theander TG et al. Antileishmanial activity of licochalcone A in mice infected with Leishmania major and in hamsters infected with Lesihmania donovani. Antimicrob Agents Chemother 1994; 38:1339-1344.

46. Piñero J, Temporal RM, Silva-Goncalves AJ et al. New administration model of transchalcone biodegradable polymers for the treatment of experimental leishmaniasis. Acta Tropica 2006; 98:59-65.

47. Akedengue B, Ngou-Milama E, Laurens A et al. Recent advances in the fight against leishmaniasis with natural products. Parasite 1999; 6:3-8.

48. Corona MC, Croft SL, Phillipson JD. Natural products as sources of antiprotozoal drugs. Curr Opin Anti-infect Invest Drugs 2000; 2:47-62.

49. Chan-Bacab MJ, Pena-Rodriguez LM. Plant natural products with leishmanicidal activity. Nat Prod Rep 2001; 18:674-688.

50. Rocha LG, Almeida JRGS, Maĉedo RO et al. A review of natural products with antileishmanial activity. Phytomed: inter j phytother phytophar 2005; 12(6-7):514-535.

51. Iwu MM, Jackson JE, Schuster BG. Medicinal plants in the figth against leishmaniasis. Parasitol Today 1994; 10(2):65-68.

52. Borris RP, Schaeffer JM. In: Nigg HN, Seigler D, eds. Phytochemical Resources for Medicine and Agriculture. New York: Plenum Press, 1992:117-158.

53. Fournet A, Gantier JC, Gautheret A et al. The activity of 2-substituted quinoline alkaloids in BALB/c mice infected with Leishmania donovani. J Antimicrob Chemother 1994; 33(3):537-544.

54. Fournet A, Hocquemiller R, Gantier J. Control of leishmaniasis. An ethno-pharmacological investigation in Bolivia. La Recherche 1995; 26(275):424-429.

55. Mercer EI. Inhibitors of sterols biosynthesis and their applications. Prog Lipid Res 1993; 4:357-416.

56. Goad LJ, Holz GG, Beach DH. Effect of the allilamine antifungal drug SF86-327 on the growth and sterol synthesis of Leishmania mexicana mexicana promastigotes. Biochem Pharmacol 1985; 34:3785-3786.

57. Beach DH, Goad LJ, Berman JD. Effects of a squalene-2,3-epoxidase inhibitor on propagation and sterol biosynthesis of Leishmania promastigotes and amastigotes. In: Hart DT, ed. Leishmaniasis. New York: Plenum, 1989.

58. Vannier-Santos MA, Urbina JA, Martiny A et al. Alterations induced by the antifungal compounds ketoconazole and terbinafine in Leishmania. J Eukaryot Microbiol 1995; 42:337-346.

59. Rangel H, Dagger F, Hernandez A et al. Naturally azole-resistant Leishmania brasiliensis promastigotes are rendered susceptible in the presence of terbinafine: compartive study with azole-susceptible Leishmania mexicana promastigotes. Antimicrob Agents Chemother 1996; 40:2785-2791.

60. Berman JD, Holz GG, Beach DH. Effects of ketoconazole on the growth and sterol biosynthesis of Leishmania promastigotes in culture. Mol Biochem Parasitol 1984; 12:1-13.

61. Goad LJ, Holz GG, Beach DH. Sterols of ketoconazole inhibited Leishmania mexicana mexicana promastigotes. Mol Biochem Parasitol 1985; 15:257-279.

62. Berman JD, Goad LJ, Beach DH et al. Effects of ketoconazole on sterol biosynthesis by Leishmania mexicana mexicana amastigotes in murine macrophage tumor cells. Mol Biochem Parasitol 1986; 20:85-92.

63. Beach DH, Goad LJ, Berman JD. Effects of lanosterol 14α-demethylation inhibitors on propagation and sterol biosynthesis of Leishmania promastigotes and amastigotes. In: Hart DT, ed. Leishmaniasis. New York: Plenum, 1989:765-771.

64. Beach DH, Goad LJ, Holz GG. Effects of antimicotic azoles on growth and sterols biosynthesis of Leishmania promastigotes. Mol Biochem Parasitol 1988; 31:141-162.

65. Kubba R, Al-Gindan Y, El-Hassan AM et al. Ketoconazole in cutaneous leishmaniasis: results of a pilot study. Saudi Med J 1986; 7:596-604.

66. Berman JD. Human leishmaniasis: clinical, diagnostic and chemotherapeutic developments in the last years. Clin Infect Dis 1997; 24:684-703.

67. Visbal G, San-Blas G, Murgich J et al. Paracoccidioides braziliensis, paracoccidioidomycosis and antifungal antibiotics. Curr Drug Targets: Infect Disord 2005; 5:211-226.

68. Haughan PA, Chance ML, Goad LJ. Effects of the azasterol inhibitor of sterol 24-transmethylation on sterol biosynthesis and growth of Leishmania donovani promastigotes. Biochem 1995; 308:31-38.

69. Magaraci F, Jimenez Jimenez C, Rodrigues C et al. Azasterols as inhibitors of sterol 24-methyltransferase in Leishmania species and Trypanosoma cruzi. J Med Chem 2003; 46:4714-4727.

70. Lorente SO, Jimenez Jimenez C, Gros L et al. Preparation of transition-state analogues of sterol 24-methyl transferase as potential anti-parasitics. Bioorg Med Chem 2005; 13:5435-5453.

71. Farrell N. Transition metal complexes as drugs and chemotherapeutic agents. In: James BR, Ugo R, eds. Catalysis by Metal Complexes Series. Vol. 11. Dordrecht: Kluwer Academic Publishers, 1989.

72. Farrell N. Uses of Inorganic Chemistry in Medicine. Cornwall: Publishers Royal Society of Chemistry, 1999.

73. Lippert B. Platinum nucleobase chemistry. Prog Inorg Chem 1989; 37:1-97.

74. Bruhn SL, Toney JH, Lippard SJ. Biological Processing of DNA modified by platinum compounds. Prog Inorg Chem 1990; 38:477-516.

75. Jamieson ER, Lippard SJ. Structure, recognition and processing of cis-platin-DNA adducts. Chem Rev 1999; 99:2467-2498.

76. Clarke MJ, Zhu F, Frasca DR. Non platinum chemotherapeutic metallopharmaceuticals. Chem Rev 1999; 99:2511-2533.

77. Guo Z, Sadler PJ. Metals in medicine. Angew Chem Int Ed 1999; 38:1512-1531.

78. Guo Z, Sadler PJ. Medicinal inorganic chemistry. Adv Inorg Chem 2000; 49:183-306.

79. Clarke MJ. Ruthenium metallopharmaceuticals. Coord Chem Rev 2002; 232:69-93.

80. Farrell NP, Williamson J, McLaren DJ. Trypanocidal and antitumour activity of platinum-metal and platinum-metal-drug dual-function complexes. Biochem pharmacol 1984; 33(7):961-971.

81. Mesa-Valle CM, Moraleda-Lindez V. Antileishmanial action of organometallic complexes of Pt (II) and Rh(I). Mem Inst Oswaldo Cruz, 1996; 9:625-633.

82. Mbongo N, Loiseau PM, Lawrence F et al. In vitro sensitivity of Leishmania donovani to organometallic derivatives of pentamidine. Parasitol Res 1997; 83:515-517.

83. Loiseau PM, Mbongo N, Bories C et al. In vivo antileishmanial action of Ir-(COT)-pentamidine tetraphenylborate on Leishmania donovani and Leishmania major mouse models. Parasite 2000; 7:103-108.

84. Mbongo N, Loiseau PM, Craciunescu DG et al. Synergistic effect of Ir-(COT)-pentamidine alizarin red and pentamidine, amphotericin B and paromomycin on Leishmania donovani. Acta Tropica 1998; 70:239-245.

85. Croft SL, Neal RA, Craciunescu DG et al. The activity of platinum, iridium and rhodium drug complexes against Leishmania donovani. Trop Med Parasitol 1992; 43:24-28.

86. Mesa-Valle CM, Moraleda-Lindez V, Craciunescu DG et al. In vitro action of new organometallic compounds against trypanosomatidae protozoa. Arzneim-Forsch 1993; 43:1010-1013.
87. Castilla JJ, Mesa-Valle CM, Sanchez-Moreno M et al. In vitro and biochemical effectiveness of new organometallic complexes of osmium (III) against Leishmania donovani and Trypanosoma cruzi. Arzneim-Forsch 1996; 46:990-996.
88. Sharquie KE, Najim RA, Farjou IB et al. Oral zinc sulphate in the treatment of acute cutaneous leishmaniasis. Clin Exp Dermatol 2001; 26:21-26.
89. Lowe G, Droz AS, Vilaivan T et al. Cytotoxicity of (2,2':6,2"-terpyridine) platinum (II) complexes to Leishmania donovani, Trypanosoma cruzi and Trypanosoma brucei. J Med Chem 1999; 42:999-1006.
90. Nguewa PA, Fuertes MA, Iborra S et al. Water soluble cationic trans-platinum complexes which induce programmed cell death in the protozoan parasite Leishmania infantum. J Inog Biochem 2005; 99:727-736.
91. Lerman LS. Structural considerations in the interaction of DNA and acridines. J Mol Biol 1961; 3:18-30.
92. Jannette KW, Lippard SJ, Vassiliades GA et al. Metallointercalation reagents. 2-hydroxyethanethiolato (2,2', 2"-terpyridine)-platinum (II) monocation binds strongly to DNA by intercalation. Proc Natl Acad Sci USA 1974; 71:3839-3843.
93. Baginski M, Fogolari F, Briggs JM. Electrostatic and non-electrostatic contributions to the binding free energies of anthracycline antibiotics to DNA. J Mol Biol 1997; 274:253-267.
94. Shui X, Peek ME, Lipscomb LA et al. Effects of cationic charge on three-dimension structures of intercalative complexes: structure of a bis-intercalated DNA complexes solved by MAD phasing. Curr Med Chem 2000; 7:59-71.
95. Suh D, Chaires JB. Criteria for the mode of binding of DNA binding agents. Bioorg Med Chem 1995; 3;723-728.
96. LePec JB, Paoletti C. A fluorescent complex between Ethidium bromide and nucleic acids. J Mol Biol 1967; 27:87-106.
97. Hogan M, Dattagupta N, Crothers DM. Transient electric dichroism studies of the structure of the DNA complexes with intercalated drugs. Biochemistry 1979; 18:280-288.
98. Wettig SH, Wood DO, Aich P et al. M-DNA: a novel metal ion complex of DNA studied by fluorescence techniques. J Inog Biochem 2005; 99:2093-2101.
99. MacArthur MW, Drisco PC, Thornton JM. NMR and crystallography complementary approaches to structure determination. Trends Biotechnol 1994; 12:149-153.
100. Holmes RJ, McKeage MJ, Murray V et al. Cis-dichloroplatinum (II) complexes tethered to 9-amino-acridine-4-carboxamides: synthesis and actino in resistant cell lines in vitro. J Inorg Biochem 2001; 85:209-217.
101. Kinnamon KE, Steck EA, Rane DS. Activity of antitumor drugs against African trypanosomes. Antimicrob Agents Chemother 1979; 15:157-160.
102. Navarro M, Cisneros-Fajardo EJ, Sierraalta A et al. Design of copper DNA intercalators with leishmanicidal activity. J Biol Inorg Chem 2003; 8:401-408.
103. Navarro M, Cisneros-Fajardo EJ, Fernández-Mestre M et al. Synthesis, characterization, DNA binding study and biological activity against Leishmania mexicana of [Cu(dppz)2]BF$_4$ J Inorg Biochem 2003; 97:364-369.
104. Navarro M, Cisneros-Fajardo EJ, Marchán E. New silver polypyridyl complexes: synthesis, characterization and biological activity on Leishmania (L) mexicana. Arzneim-Forsch 2006; 56:600-604.
105. Navarro M, Hernández C, Colmenares I et al. Synthesis and characterization of [Au(dppz)$_2$]Cl$_3$. DNA interaction studies and biological activity against Leishmania (L) mexicana. J Inorg Biochem 2007; 101:111-116.
106. Long EC, Barton KJ. On demonstrating DNA intercalation. Acc Chem Res 1990; 23:273-279.
107. Haq I, Lincoln P, Suh D, et al. Interaction of Δ- and Λ-[Ru(phen)$_2$DPPZ]$^{2+}$ with DNA: A calorimetric and equilibrium binding study. J Am Chem Soc 1995; 117:4788-4796.
108. Xiong Y, Ji LN. Synthesis, DNA binding and DNA-mediated luminescence quenching of Ru(II) polypyridine complexes. Coord Chem Rev 1999;185-186: 711-733.

CHAPTER 7

Programmed Cell Death during Malaria Parasite Infection of the Vertebrate Host and Mosquito Vector

Luke A. Baton, Emma Warr, Seth A. Hoffman and George Dimopoulos*

Abstract

In recent years, there has been an increasing awareness of the role of programmed cell death (PCD) in the malaria parasite's infection of its vertebrate host and mosquito vector. Although the evidence that PCD occurs within malaria parasites themselves is currently limited and controversial, a significant body of research now indicates that PCD of both vertebrate host and mosquito vector cells plays an important, if still incompletely understood, role during infection with this parasite. A greater understanding of the role of PCD during malaria infection of the vertebrate host and mosquito vector may lead to the development of novel intervention strategies that can reduce the burden of the disease. Here we review the current evidence for the existence of PCD within malaria parasites themselves and discuss the recent fascinating advances in our understanding of the occurrence of PCD in vertebrate host and mosquito vector cells during malaria infection.

Introduction

Malaria is caused by parasitic unicellular eukaryotes belonging to the apicomplexan genus *Plasmodium*. There are many different species of malaria infecting mammals, birds and reptiles.[1] Four species of *Plasmodium* commonly infect humans, the most prevalent and severe form being *P. falciparum*, which is found throughout the sub-tropical and tropical regions of the world.[2,3] Malaria causes an estimated 300 to 660 million clinical cases of infection and 1 to 3 million deaths, annually.[3] Approximately 70% of all human malaria infection occurs in Africa, with infants and children under the age of 5 years bearing the main burden of disease.[3] Despite the World Health Organization-sponsored program to "Roll Back" malaria and reduce the global incidence of the disease by 2010, a number of complex biological, cultural, economic, political and social factors have caused a resurgence of malaria in recent years.[4] Current control methods are failing because of the emergence of drug resistance in the parasites[5] and insecticide resistance in their mosquito vectors,[6] and even after several decades of intensive research, no effective and practical long-term vaccine is yet available.[4] It is hoped that a greater understanding of the basic

*Corresponding Author: George Dimopoulos—W. Harry Feinstone Department of Molecular Microbiology and Immunology, Malaria Research Institute, Bloomberg School of Public Health, Johns Hopkins University, 615 N. Wolfe Street, Baltimore MD 21205-2179, USA. Email: gdimopou@jhsph.edu

Programmed Cell Death in Protozoa, edited by José Manuel Pérez Martín.
©2008 Landes Bioscience and Springer Science+Business Media.

biology of *Plasmodium* in its vertebrate host and mosquito vectors, including an understanding of the role of PCD in the interaction between these organisms, will lead to the development of novel intervention strategies.

The Life Cycle of *Plasmodium*

Malaria parasites undergo a complex life cycle involving development within a vertebrate host and subsequent transmission between these hosts by hematophagous invertebrate vectors, typically mosquitoes (see refs. 7-9 and their associated websites for detailed descriptions and computer animations of the parasite's life cycle).

Plasmodium species infecting humans are transmitted by various mosquito species belonging to the genus *Anopheles*. Infection of the vertebrate host is initiated when sporozoite-stage malaria parasites are inoculated into the dermis during blood-feeding by infected female mosquitoes.[10,11] The sporozoites migrate to the liver through the bloodstream, or possibly lymphatics,[12] and invade the liver parenchyma, where intracellular development occurs within hepatocytes.[13] Within the hepatocyte, the parasite grows in size and undergoes multiple rounds of asexual multiplication, culminating in the release of many thousands of merozoite-stage parasites into the bloodstream.[9] Each merozoite rapidly invades an erythrocyte via the formation of a parasitophorous vacuole, which continues to surround the parasite throughout intra-erythrocytic development.[7] Like the sporozoites within hepatocytes, the merozoites grow in size and undergo asexual multiplication, culminating in the release of further daughter merozoites, which invade new erythrocytes and repeat the cycle of intra-erythrocytic development.[7] However, a small proportion of merozoites undergo sexual differentiation, developing into transmission stages known as gametocytes, which then transform within the mosquito vector after ingestion during blood-feeding.[8] Once within the mosquito vector, the gametocytes undergo a complex series of transformations, including the formation of motile invasive stages known as ookinetes, which migrate through the ingested blood and invade the surrounding epithelium of the mosquito midgut.[8] Malaria parasite development in the mosquito culminates in the accumulation of sporozoites in the mosquito salivary glands, ready for inoculation into a new vertebrate host.[8]

What Are PCD and Apoptosis?

Before describing the evidence for PCD in the malaria parasites themselves, it is important to note that observing an apoptotic phenotype is not equivalent to demonstrating the existence of PCD. Conversely, the absence of an apoptotic phenotype is not equivalent to demonstrating the absence of PCD. This distinction is especially important in the case of unicellular eukaryotes, in which the functional significance of apoptosis has not usually been experimentally investigated, let alone established. According to Ameisen's[14-16] hypotheses about the evolutionary origins and significance of apoptosis, at least in the beginning, there were no bona fide genetic cell death programs, only programs for various processes enabling life, such as metabolism, cell cycle progression and cell differentiation. From these processes there evolved mechanisms to prevent their dysregulation and these processes subsequently became "exapted" into the pathways of PCD seen in metazoa today. In this view, the distinction between apoptosis and necrosis becomes blurred and it is currently uncertain where on the evolutionary continuum protozoa (and PCD) are located. Conceivably, apoptotic phenotypes might be characteristic of cells that lack evolved PCD pathways and that die "necrotically" through an accumulation of errors in the cellular machinery responsible for processes such as metabolism and cell cycle progression. This perspective is further reinforced by the discovery of caspase-independent pathways of PCD with "necrotic" phenotypes (for a deconstruction of the apoptosis/necrosis dichotomy see Broker et al[17]). In this review, we refer to PCD throughout as an adaptive mechanism of cell suicide that has evolved through natural selection, while we take apoptosis to mean a set of conventional phenotypic characteristics possessed by some dying cells that may, or may not, be indicative of PCD.

PCD within Malaria Parasites Themselves

Theoretical Considerations: Why Would PCD Exist within Malaria Parasites?

In metazoa, PCD is traditionally perceived as a homeostatic mechanism that maintains the integrity of the whole organism by enabling the removal of unwanted healthy cells during normal development and growth processes, together with the destruction of unhealthy damaged, precancerous and/or pathogen-infected cells whose continued existence threatens the well-being of the organism.[18-20] According to this view, individual cells of a particular multicellular organism "altruistically" commit suicide for the benefit of the group of genetically-related cells that comprise the entire multicellular organism. Consequently, the requirement for PCD within unicellular organisms is not immediately apparent.[15] PCD in a unicellular organism is seemingly a one-way ticket to rapid extinction and it is not obvious how PCD could evolve in unicellular organisms since, in terms of propagating "selfish genes," such a trait would apparently represent genetic suicide.[15]

However, many unicellular organisms, including protozoa such as malaria parasites, can be viewed as being part of a "virtual" multicellular organism[15,21] whose constituent cells are physically separated and distributed throughout space and time, rather than being immediately adjacent to one another as in the "conventional" multicellular organism. For example, malaria infections of vertebrate hosts can be initiated by a small number of genetically identical sporozoites. Consequently, all the asexual and sexual erythrocytic stage parasites resulting from these sporozoites will also be genetically identical; such a circumstance is referred to as a clonal infection. Indeed, individual genetically identical asexual erythrocytic stage parasites found within a single vertebrate host can be viewed as equivalent to the somatic cells of a multicellular organism, while the sexual erythrocytic stage parasites—the gametocytes—are analogous to the germ cells of multicellular organisms. Under circumstances of clonal infection, kin selection[22] provides a mechanism whereby PCD can evolve within unicellular organisms, with the "altruistic" act of suicide by some individual malaria parasites benefiting the reproductive success of genetically related individuals present within the same host. If PCD by some individual malaria parasites within a clonal infection enhances transmission to new hosts of genetically related malaria parasites that do not undergo PCD, then it is possible for PCD to evolve and spread throughout the population of malaria parasites. A similar argument also applies to the malaria parasite stages that infect the invertebrate host, as mosquito vectors can ingest genetically identical gametocytes resulting from a clonal infection of the vertebrate host.

The preceding argument holds true when malaria parasite infections are clonal and genetically identical (or even just closely related) parasites are found together within the same host and/or vector. However, it should be noted that malaria parasite infections are frequently not clonal but are rather composed of multiple, genetically distinct parasite clones[23-25] that can each be conceived of as a distinct multicellular organism. In fact, the number of malaria parasite clones infecting vertebrate and invertebrate hosts exhibits a positive, although not necessarily linear, relationship with the intensity of malaria transmission: where transmission is low, most infections are clonal, but where transmission is high, infections tend to be multi-clonal.[26-30] Under circumstances of multiple-clone infection, the evolutionary outcome is uncertain and complex because the altruistic acts of PCD committed by one malaria parasite clone can potentially be "exploited" by genetically different malaria parasite clone(s) that happen to be infecting the same vertebrate host or mosquito vector. The second malaria parasite clone(s) can "cheat" and receive the benefit of PCD produced by the first, without paying any of the costs associated with the altruistic PCD behavior (such as reduced growth of parasite numbers).

In general, evolutionary theory predicts that altruistic behaviors such as PCD will tend to be unstable and be selected against in circumstances in which they are susceptible to cheating, although the exact outcome very much depends on the particular costs and benefits of the trait in question, as well as the exact details of the population structure of the relevant organism. Consequently, whether or not PCD is found in malaria parasites may (or may not!) vary according to the level of malaria transmission. These arguments are further complicated by the possibility of selection occurring simultaneously, at different levels, in different directions (i.e., within and between hosts/vectors). For detailed discussions and examples of the evolutionary logic underlying these arguments, as well as the importance of population structure in the evolution of altruistic traits, the reader is referred to references 31-34.

Given that the conditions necessary for the evolution of altruistic PCD might exist for malaria parasites (at least in areas of relatively low transmission, where clonal infections predominate), it is possible to speculate on the different ways in which PCD might benefit these parasites during the infection of their vertebrate and invertebrate hosts. Two obvious scenarios are as a mechanism for (1) modulating the host's immune responses (through reducing inflammatory responses elicited as a consequence of extracellular release of cytosolic contents and/or restricting antigen availability/presentation to cells of the host immune system, although it is currently unclear how and whether such a mechanism would work); and (2) regulating the density of malaria parasite stages found within the host (and thus preventing premature and/or "over-exploitation" of the host, i.e., death of the host before parasite transmission to a new host can occur).[35] In the latter scenario, malaria parasites would presumably require a mechanism for sensing, either directly or indirectly, the density of other malaria parasites present within the host and responding appropriately. There is already evidence that malaria parasites can respond facultatively to the environment of the host bloodstream during sexual differentiation (reviewed in ref. 36 among others) and investigations into the possibility of communication between erythrocytic stage malaria parasites have been undertaken.[37]

Another proposed benefit of PCD to unicellular organisms is selection of the "fittest cells" and removal of developmentally defective cells from the clonal population.[15,35] Unlike multicellular organisms in which cellular specialization makes the fate of each cell within the organism dependent upon the fate of the others, it is not obvious that the continued existence of developmentally defective cells threatens the future survival of unicellular organisms such as malaria parasites, whose relative lack of specialization enables independent existence from other members of the clonal population (cf., with cancer in metazoan organisms). However, the important issue here is perhaps not specialization per se but the notion of "shared fate," and the extent to which the presence of developmentally defective (or just superfluous) individual parasites alters the environment for other members of the clone and prevents their continued survival and reproduction. If the presence of developmentally defective or superfluous parasites does not adversely alter the environment for fellow clone members, then there is no selection pressure to drive the evolution of PCD in this context.

The possibility should also be considered that PCD might not be beneficial for the malaria parasites themselves (see refs. 14-16 for the example of toxin-antidote modules in bacteria). Induction of PCD within malaria parasites could be used by the host as a defense mechanism to control and limit the level of infection, although this scenario would need to explain why mechanisms of PCD exist within the parasites that can be exploited by the host.[38]

Practical Considerations: How Would Malaria Parasites Commit PCD?

Although there are theoretical grounds for believing that PCD could evolve within malaria parasites themselves, there is currently little evidence that this phenomenon actually occurs within malaria parasites. The evidence that is currently available is contradictory and ambivalent because it is not clear that the described "apoptotic" morphologies are genuinely indicative of an evolved mechanism of PCD, as opposed to non-adaptive forms of cell death that incidentally, or because of common ancestry of the molecular mechanisms involved, display the biochemical and morphological characteristics of apoptosis in metazoa.

One way to confirm the existence of PCD within malaria parasites is to demonstrate the presence of the molecular machinery for PCD within these organisms, although there is no guarantee that homologues present in different species have the same functions. However, searches of the *Plasmodium* genome have failed to identify homologues of many of the classical molecular components of the extrinsic and intrinsic PCD pathways in metazoa (reviewed in refs. 35,39). The effector caspases regulating PCD in higher eukaryotes have not been found, although a single metacaspase, a single calpain and numerous other uncharacterized cysteine proteases have been identified in *P. falciparum*.[35,40-42] Several mitogen-activated protein kinases (MAPK), which are implicated in caspase-independent forms of PCD in metazoa,[43] have also been identified in *Plasmodium*.[44] However, homologues of apoptosis-inducing factor (AIF), p53, Bcl-2, endonuclease G and apoptotic protease activating factor 1 (Apaf-1) have been sought but not found in malaria parasites.[35] The significance of this apparent absence of many of the molecular components of the classical metazoan PCD pathways is currently uncertain: It may reflect a genuine absence of these pathways (and PCD) in *Plasmodium,* or the components of these pathways may be sufficiently divergent in malaria parasites that standard BLAST searches are unable to detect them. A further possibility is that PCD in malaria parasites uses different molecular components/pathways and/or is fundamentally different from the process operating in metazoa.

Evidence for PCD in Malaria Parasite Stages Occurring in the Vertebrate Host

There is currently no decisive published evidence that PCD occurs within the malaria parasite stages that infect the vertebrate host.[35] The most convincing evidence for PCD within malaria parasites themselves comes from studies on the parasite's development within host hepatocytes. van de Sand et al[41] have reported the occasional presence of TUNEL-positive nuclei in malaria parasites, within infected hepatocytes that also possessed TUNEL-positive nuclei. These authors interpreted this observation as evidence that malaria parasites occasionally die (for unknown reasons) within their host hepatocytes; in dying, the malaria parasites would cease production of the anti-apoptotic factors that had previously prevented the host hepatocytes from undergoing PCD and hepatocyte PCD would then occur after the death of the infecting parasites. (See below for further details of parasite-mediated inhibition of host hepatocyte PCD and its possible role in immunomodulation of the host's response to malaria infection.) Again, the significance of this observation is unknown: Is it evidence of an evolved mechanism of malaria parasite PCD or just non-adaptive, "necrotic" cell death induced by host immune responses and/or developmental defects within the malaria parasite itself? If malaria parasites within the liver are genuinely undergoing PCD, further research is clearly required to understand why this process is happening and to determine its significance for the host-parasite interaction.

There have also been suggestions that PCD takes place in the asexual erythrocytic stages within the vertebrate host.[35,45] Picot et al[45] have speculated that "crisis forms" (erythrocytic malaria parasite stages with stress-induced abnormal morphology that often appear both in vivo and in vitro during periods of high parasitemia and/or immunological responses) undergo apoptosis. They used agarose gel electrophoresis to show that the DNA of a *P. falciparum* clone sensitive to the drug chloroquine was degraded (as evidenced by internucleosomal DNA fragmentation) following treatment with this antimalarial compound, while the DNA of a chloroquine-resistant *P. falciparum* clone remained intact. The significance of this observation in the absence of morphological studies is somewhat doubtful, as the DNA degradation observed in the chloroquine-sensitive clone arguably reflected necrosis rather than PCD. Deponte and Becker[35] have also described unpublished work that has demonstrated that various (unspecified) antimalarial drugs induce apoptosis in *P. falciparum,* as exemplified by the occurrence of TUNEL-positive nuclei in asexual erythrocytic stages of this malaria parasite following in vitro drug treatment.

From an evolutionary perspective, it seems doubtful that PCD would occur as an adaptive response by the malaria parasite to antimalarial treatment per se, although it is certainly within reason that various antimalarial compounds may trigger apoptotic processes that are either non-adaptive for the malaria parasites themselves or have evolved for other reasons. However, recent research by Nyakeriga et al[46] on the effects of chloroquine and other antimalarial compounds on the asexual erythrocytic stages of drug-sensitive *P. falciparum* failed to find any biochemical or morphological evidence that these compounds induced apoptosis in these malaria parasite stages: Laddering typical of internucleosomal DNA fragmentation/chromatin condensation was not observed by gel electrophoresis, nor were TUNEL-positive parasite nuclei identified. Furthermore, there was no evidence that chloroquine- or atovaquone-treated malaria parasites possessed apoptotic-like or "crisis form" morphology. Rather, malaria parasites treated with these two drugs appear to undergo arrested development at defined points during the asexual erythrocytic developmental cycle. The latter study also showed that chloroquine-treated malaria parasites did not have altered mitochondrial membrane potential, while those treated with the antimalarial compound atovaquone did.[46] However, atovaquone is an inhibitor of cytochrome *c*, a part of the mitochondrial electron transport chain and a key player in the induction of the intrinsic pathway of PCD in metazoans. Therefore, the significance of this observation as evidence for the existence of PCD in the asexual erythrocytic stages of malaria parasites is questionable.

Nyakeriga et al[46] have concluded that if PCD occurs in malaria parasites, it does not have the features typically associated with apoptosis in multicellular eukaryotes. Similarly, Pankova-Kohlmyansky et al[47] also failed to find any evidence that the antimalarial drugs artemisinin and mefloquine induce apoptosis in the asexual erythrocytic stages of *P. falciparum*. These authors showed that the antimalarial activity of artemisinin and mefloquine in vitro is dependent on the presence of reduced levels of malaria parasite glutathione, resulting from the production of ceramide mediated by endogenous sphingomyelinase activity. As ceramide is implicated as a pro-apoptotic signal in some systems,[48] Pankova-Kohlmyansky et al[46] investigated whether artemisinin and mefloquine could induce apoptosis in the asexual erythrocytic stages of *P. falciparum*. However, they saw no evidence of DNA fragmentation, TUNEL-positive nuclei, or "caspase-3" activation after exposure to these drugs. Furthermore, electron microscopy failed to reveal any signs of apoptotic morphology in treated *P. falciparum* cultures.[47]

Although there is currently little published evidence that antimalarial drugs induce either PCD or apoptosis, there is some suggestion that oxidative stress may induce apoptosis within malaria parasites. TUNEL-positive nuclei have been reported in the asexual erythrocytic stages of *P. falciparum* after hydrogen peroxide (H_2O_2) treatment,[35] and the nitric oxide (NO) donor S-nitroso-N-acetyl-D, L-penicillamine (SNAP) has been shown to induce "crisis form" morphology and minor DNA degradation; these findings may indicate that host immune responses trigger PCD in these stages of the malaria parasite.[46] However, in the latter study, the DNA degradation observed did not resemble the DNA laddering typical of apoptotic mammalian cells and TUNEL-positive nuclei were not found.[46] Infection-induced NO production by hepatocytes has also been proposed to account for the occurrence of TUNEL-positive malaria parasites within these host cells,[41] although this hypothesis has yet to be formally tested.

Evidence for PCD within Malaria Parasite Stages Occurring in the Mosquito Host

The strongest evidence for the occurrence of PCD within malaria parasites themselves comes from studies of the ookinete stage that occurs within the mosquito midgut lumen.[49,50] Al-Olayan et al[49,50] have shown both in vitro and in vivo that ookinetes of the rodent malaria parasite *P. berghei* exhibit a number of morphological and biochemical characteristics of apoptosis, including condensed and TUNEL-positive nuclei, annexin V-positivity indicating phosphatidylserine externalization and loss of membrane impermeability (as measured by propidium iodide and trypan blue staining). These authors also demonstrated the presence of "caspase" activity in the cytoplasm of the ookinetes using a carboxyfluorescein-labeled caspase inhibitor probe and were able to reduce the proportion of ookinetes undergoing apoptosis by the addition of various caspase inhibitors, including Z-VAD.fmk and Boc-ASP.[49,50] A comparison of different caspase inhibitors further showed that ookinete apoptosis was reduced by the caspase-3 subfamily inhibitor Z-DEVD.fmk but not by the caspase-1 subfamily inhibitor Z-YVAD.cmk. Western blot analysis of ookinete cell extracts using biotin-VAD identified two bands of approximately 45 and 28 kDa whose identity remains unknown. With regard to the existence of PCD (rather than just apoptosis) within malaria parasites, perhaps the most important observation made by these authors was that the addition of

Z-VAD.fmk to an infectious gametocyte-containing bloodmeal significantly increased the number of oocysts that formed in *A. stephensi*.[49] This experiment needs to be interpreted with care because mosquito midgut cells invaded by the ookinete also undergo apoptosis, as discussed below and the presence of Z-VAD.fmk may have altered this process as well. However, the results suggest that ookinetes undergoing apoptosis are not simply "necrotic" or otherwise developmentally defective, as they are apparently able to complete subsequent development, at least to the oocyst stage, if rescued by the presence of a caspase inhibitor. Al-Olayan et al[49] have speculated that ookinete apoptosis may act as a PCD mechanism for limiting the level of malaria parasite infection within the mosquito vector and thus for enhancing transmission between vertebrate hosts by extending mosquito longevity, but no evidence was presented that the occurrence of apoptosis within ookinetes is density-dependent, as might be predicted under such a hypothesis.

PCD of Vertebrate Host Cells Associated with Malaria Infection

In contrast to the current lack of clear-cut evidence for the existence of PCD within malaria parasites themselves, there is a growing body of research demonstrating the occurrence and importance of PCD in the cells of the hosts infected with the parasite. Although there is little doubt that host cells undergo (or are prevented from undergoing) apoptosis following malaria infection, the adaptive significance of these observations is not necessarily always clear (i.e., whether the observed phenomenon benefits the host or the parasite, or both or neither and whether such phenomena should be regarded as adaptations on the part of the host and/or malaria parasite).

Sporozoite Infection of the Liver and Subsequent Development within Hepatocytes

Several studies have reported that malaria parasites developing intracellularly inhibit the PCD of their host hepatocytes.[41,51] The tremendous intracellular growth of the malaria parasite greatly distorts the size and shape of the infected host hepatocyte, a stressor that would normally be expected to induce apoptosis.[9,41] However, most *P. berghei*-infected hepatocytes have morphologically normal nuclei and mitochondria, exhibit neither caspase activation nor differential expression of the apoptosis-related markers Bax and Bcl and possess a reduced sensitivity to a variety of inducers of PCD, including the drug cycloheximide, tumor necrosis factor, D-galactosamine, oxidative stress (peroxidase treatment), UV radiation and serum deprivation. Different mechanisms of host cell PCD inhibition seem to operate during the early and late stages of hepatocyte infection. Carrolo et al[52] initially found evidence that intracellular migration of sporozoites through the liver resulted in the release of hepatocyte growth factor (HGF) from the traversed liver cells and this HGF rendered the surrounding uninvaded hepatocytes susceptible to sporozoite infection, an effect that was mediated by the HGF receptor, MET. Further studies by Leiriao et al[51] have shown that pretreatment of hepatocytes with HGF, or binding of its receptor MET with specific antibodies, promoted parasite development within hepatocytes and reduced host cell apoptosis, while repression of MET through RNAi or heterologous expression of a chimeric dominant interfering protein has the converse effect. These observations have demonstrated a role for HGF-MET signaling in preventing hepatocyte PCD during the early stages of parasite establishment within the host cells. Accordingly, inhibition of the PI3-kinase/Akt signal transduction pathway, which mediates the anti-apoptotic activity of HGF-MET signaling in other systems, results in an increase in the apoptosis of *P. berghei*-infected hepatocytes within the first 6 hours after infection.[51] In contrast, van de Sand et al[41] have reported that inhibition of the PI3-kinase/Akt transduction pathway has no effect on the level of apoptosis exhibited by malaria-infected hepatocytes at 24 and 48 hours post-infection, suggesting that HGF-MET signaling is not responsible for the elevated levels of resistance to apoptosis that are observed in hepatocytes during the later stages of parasitic infection.

However, some malaria-infected hepatocytes do undergo apoptosis, exhibiting the characteristic changes in nuclear morphology that include TUNEL positivity, nuclear translocation of relA (NF-κB p65) and caspase-3 activation.[41,51,53] The reason that some malaria-infected hepatocytes undergo apoptosis and others infected with apparently genetically identical sporozoites from the same clone do not is currently unknown. As discussed above, at least in some instances, hepatocyte apoptosis is preceded by the death of the infecting malaria parasite.[41] Regardless of the cause, it has become increasingly apparent that apoptosis of malaria-infected hepatocytes may play an important role in the acquisition of protective immune responses by the host (although this interpretation is contested—see refs. 54-56 for a discussion). For example, malaria-infected hepatocytes undergoing apoptosis in vivo are surrounded by lymphocytes and are phagocytosed, whereas viable parasites in non-apoptotic hepatocytes do not induce local inflammatory responses.[41,53] Furthermore, apoptosis is also undergone by the host cells of genetically- or radiation-attenuated malaria parasites that are able to infect but not complete development within hepatocytes, suggesting that the protective effect of experimental vaccines comprised of these developmentally defective parasites is a consequence of this phenotype.[53,57] These studies have also provided further evidence that inhibition of PCD of host hepatocytes is the result of active suppression by infecting malaria parasites.

As is true for other related apicomplexan parasites that also develop intracellularly and inhibit host cell apoptosis, such as *Theileria* and *Toxoplasma,* malaria parasites have presumably evolved the means to manipulate the pathways regulating PCD in their host cells (reviewed in refs. 21,58,59). As yet, though, the mechanism(s) and malaria parasite molecule(s) that mediate this inhibition of host cell apoptosis have not been identified. However, P36p, a member of the *Plasmodium* genus-specific 6-cys protein family, expressed in sporozoites and of unknown function, has been implicated in the repression of host cell PCD: Hepatocytes infected with genetically-modified *P. berghei* lacking P36p exhibit higher levels of apoptosis than do hepatocytes infected with wild-type parasites.[57] Although the mechanism by which the loss of P36p triggers host cell apoptosis is not understood, it has been shown that P36p mutant parasites lack the parasitophorous vacuole that normally surrounds malaria parasites during intracellular development within hepatocytes and separates them from the host cell cytoplasm.[9] Consequently, these mutants reside directly within the host cell cytoplasm and therefore they do not benefit from the protection that is potentially conferred by the parasitophorous vacuole, which may facilitate immune evasion by shielding wild-type parasites from host cell recognition and defense mechanisms.[57]

Development of the Malaria Parasite within Erythrocytes

PCD has traditionally been thought to be absent from mature human (and other mammalian) erythrocytes because these cells lack nuclei and mitochondria and are resistant to inducers of apoptosis in nucleated mammalian cells, including serum deprivation and the protein kinase inhibitor staurosporine. However, recent studies have demonstrated that erythrocytes exhibit a PCD process known as eryptosis, which is independent of the caspase-mediated and mitochondrial pathways (reviewed in refs. 60,61). Eryptosis is induced by Ca^{2+} influx, mediated by the cysteine protease calpain and characterized by classical apoptotic hallmarks such as plasma membrane microvesiculation, cell shrinkage, phosphatidylserine externalization and subsequent rapid engulfment by macrophages.[62-65] The characteristics of eryptosis are similar to those of erythrocyte aging,[66] suggesting that this form of PCD is involved in the phagocytic removal of senescent erythrocytes from the circulation.[62,63,67]

Whether erythrocyte PCD is also activated as an evolved host defense mechanism following erythrocyte infection is currently unknown and this possibility has not been specifically investigated in malaria parasites, although the possible fascinating implications of eryptosis for malaria parasites have been discussed by Lang et al.[68] Given the manipulation of hepatocyte PCD by malaria parasites during infection of the liver, the possibility of similar manipulation of host cell PCD during both asexual and sexual erythrocytic malaria parasite development should be given serious consideration.

There is evidence that, in vitro, intra-erythrocytic stage *P. falciparum* partially induce eryptosis through an increase in nonselective Ca^{2+} uptake by the host erythrocyte, which triggers the externalization of phosphatidylserine observed in malaria-infected erythrocytes.[68,69] However, the rate of up-take of Ca^{2+} by *P. falciparum*-infected erythrocytes is apparently no greater than that of uninfected erythrocytes under conditions of glucose abundance. Therefore, it appears that parasitization does not result in a Ca^{2+} influx sufficient to trigger eryptosis under these circumstances.[70] Consequently, Ca^{2+}-independent mechanisms may also be responsible for the phosphatidylserine externalization of malaria-infected erythrocytes.[68] Under conditions of glucose deprivation, however, the uptake of Ca^{2+} by late-stage *P. falciparum*-infected erythrocytes is significantly increased, leading Staines et al[70] to speculate that glucose starvation resulting from high parasitemia could act as a density-dependent mechanism to regulate parasite numbers and prevent over-exploitation of the host.

In contrast to mammalian hosts, the erythrocytes of the reptilian and avian hosts of malaria parasites are nucleated and in chickens they are sensitive to serum deprivation and staurosporine and they undergo a caspase-independent but otherwise classical apoptosis involving nuclear disintegration.[71] Consequently, avian malaria parasites such as *P. gallinaceum*, which infects domestic fowl, might possess mechanisms that are distinct from those of mammalian malaria parasites and that allow them to inhibit PCD of their host erythrocytes.

A long-standing enigma in malaria biology is why malaria parasites undergo the process of sequestration: removal of trophozoite- and schizont-infected erythrocytes from the peripheral blood circulation through adhesion to the postcapillary vascular endothelium of various vital organs.[72] Sequestration is thought to allow malaria-infected erythrocytes to avoid passage through the spleen and therefore phagocytic clearance from the bloodstream.[72,73] Sequestration as a mechanism of immune evasion appears to be paradoxical because it is mediated by parasite molecules that are inserted into the outer membrane of the host erythrocyte and that expose the otherwise hidden, intracellular and immunoprivileged malaria parasite to the host immune system.[73] In this context, the phosphatidylserine externalization characteristic of *P. falciparum*-infected erythrocytes, which would promote phagocytic clearance within the spleen,[68] might necessitate sequestration and provide a receptor function that effectively mediates the adherence of parasitized erythrocytes to the vascular endothelium.[74] In accord with the foregoing discussion, the protective effect of various hemoglobinopathies, such as sickle-cell trait, against severe malaria disease may result from the greater propensity of erythrocytes from people with these conditions to undergo eryptosis (and, hence, for malaria-infected erythrocytes to be more readily cleared from circulation by phagocytosis).[68,75]

Pathophysiology of Malaria Infection in the Vertebrate Host

Malaria infection of the vertebrate host often results in systemic multi-organ disorders with complex pathophysiology (reviewed in refs. 76-82). In *P. falciparum* and various animal models, sequestration is thought to be a major contributory factor to the clinical symptoms of malaria disease, including the potentially fatal manifestation of cerebral malaria.[79,82] Sequestration of malaria-infected erythrocytes is mediated through the parasite receptors inserted in the infected host erythrocyte membrane. These receptors interact with various cellular adhesion molecules that belong to the immunoglobulin and integrin families expressed on the surface of host endothelial cells (reviewed in refs. 82,83). Binding of malaria-infected erythrocytes to these molecules, including ICAM1 and CD36, may trigger intracellular signaling cascades in the host endothelial cells.[84] An increasing number of studies have indicated that in vitro and in vivo adherence of malaria-infected erythrocytes specifically induces apoptosis in endothelial cells from a variety of organs, including the brain, lung and kidney.[85-90]

Pino et al[87] have shown that expression of various pro-apoptotic genes, including those encoding members of the TNF superfamily (Death receptor 6, Fas and FasL), the Bcl-2 family (*bad, bax, bcl-w, bik*) and other apoptosis-related gene products (*caspase-3, DDF45/ICAD, IFN-γ receptor 2, iNOS, SARP2*), is transcriptionally up-regulated in cultures of human lung endothelial cells following adherence of *P. falciparum*-infected erythrocytes. These authors further demonstrated the occurrence of apoptosis in endothelial cells by electron microscopy, annexin V-binding, DNA fragmentation and assays of caspase activation and inhibition.[87] Specific inhibition of caspase-8 and caspase-9 revealed that both the extrinsic death receptor-mediated and intrinsic mitochondrial pathways are involved in the apoptosis of endothelial cells and that NO-mediated oxidative stress is also implicated.[86,87]

Wassmer and coworkers[89] have further shown that the ability of malaria-infected erythrocytes to induce apoptosis in human brain endothelial cells in vitro is potentiated by prior TNF activation and binding of platelets to the endothelial cells. These authors also demonstrated that platelet-derived TGF-β is important for the pro-apoptotic effect of platelet binding.[91] Hemmer et al[85] have investigated the effect of co-incubation of human sera and naïve neutrophils on the occurrence of apoptosis in human umbilical endothelial cells. They found a positive correlation between the severity of the serum donor's malaria disease and the ability of the donor's serum to induce apoptosis in endothelial cells.

Other studies have used the *P. berghei*-based murine model to characterize the role of apoptosis in the pathophysiology of cerebral malaria. Potter et al[88] have shown that CD8[+] lymphocytes induce apoptosis of endothelial cells via a perforin-dependent process during *P. berghei* infection. These studies suggest that endothelial cell apoptosis plays an important role in the disruption of the blood-brain barrier during malaria infection and contributes to cerebral edema and inflammation (reviewed in refs. 78,84).

Although there is good evidence for apoptosis of vascular endothelial cells during cerebral malaria, the role of apoptosis in other aspects of cerebral pathology associated with malaria infection is less certain. The coma and encephalopathy accompanying cerebral malaria are sometimes reversible and individuals often appear to recover without permanent neurological complications.[79] However, in some instances, individuals recovering from cerebral malaria suffer lasting neurocognitive dysfunction.[79] For this reason, the possible occurrence of apoptotic neuronal death during malaria infection has been of some interest. Several studies have reported elevated levels of neuronal apoptosis as indicated by TUNEL assay,[90,92] while other studies using markers of caspase-3 and/or caspase-8 activation have failed to identify any increase in neuronal apoptosis in individuals with cerebral malaria.[93,94] Other brain cells may also undergo apoptosis during cerebral malaria. For example, in the *P. berghei* model of cerebral malaria, astrocytes exhibit caspase-3 activation and undergo Fas-mediated apoptosis.[95]

P. falciparum-infected erythrocytes also sequester within the blood vessels of the uterus, causing so-called "placental malaria" in pregnant women, a major cause of morbidity and mortality for both mothers and their unborn children.[76] Sequestration of *P. falciparum*-infected erythrocytes results in damage to the syncytiotrophoblast, which is probably responsible for the low neonatal birth weight of infants born to mothers who suffered malaria during pregnancy.[76] Crocker et al[96] have investigated the occurrence of apoptosis in term placentae from malaria-infected women but found no evidence for elevated levels of PCD. Rather, placental cell death in malaria-infected women appeared to result from necrosis.

Guha et al[97] have also reported that in the livers of mice infected with the rodent malaria parasite *P. yoelii,* asexual erythrocytic-stage infection results in the induction of hepatocyte apoptosis associated with oxidative stress. In this system, apoptotic liver cells from malaria-infected mice exhibited TUNEL-positive nuclei and caspse-3 activation. This activation was apparently mediated through the intrinsic mitochondrial pathway, as evidenced by transcriptional down-regulation of *bcl-2* and up-regulation of *bax,* translocation of Bax from the cytosol into the mitochondria, loss of mitochondrial membrane potential and release of cytochrome c into the cytosol. In contrast, differential transcription of fas and caspase-8 activation were not observed, suggesting that the extrinsic death receptor pathway was not responsible for the apoptosis observed within the liver of *P. yoelii*-infected mice. The oxidative stress causing apoptosis apparently resulted from increased levels of the hydroxyl radical in the livers of malaria-infected mice and the apoptosis of liver cells could be reversed by treating the mice with hydroxyl scavengers such as DMSO, mannitol and spin trap PBN.[97] It is important to note that the apoptosis observed within the liver during these experiments was not a consequence of sporozoite invasion of hepatocytes, since the infections were initiated through intraperitoneal injection of erythrocytic-stage parasites. Whether the apoptosis induced within the liver by infection with asexual erythrocytic-stage parasites is the result of sequestration and contributes to pathology during human malaria infections has yet to be determined.

Suppression of Vertebrate Host Immune Responses

Nonsterile immunity to malaria infection is gradually acquired over a prolonged period of repeated exposure and, once acquired, is short-lived and rapidly lost in the absence of re-infection (reviewed in refs. 81,98). The reasons for the failure of vertebrate hosts to acquire long-lasting, sterile, protective immunity to malaria infection are complex and imperfectly understood, but there is increasing evidence that development of the vertebrate host's immune response to current and subsequent malaria infections is modulated by *Plasmodium*-associated apoptosis of host white blood cells.

A number of studies have demonstrated elevated levels of apoptosis in peripheral blood lymphocytes, primarily T-cells, obtained during and after malaria infection.[99-106] In human infections and experimental animal models, elevated levels of T-cell apoptosis in vitro and ex vivo are correlated with *Plasmodium*-associated T-lymphocytopenia, disease severity, elevated serum levels of soluble Fas ligand and reduced numbers of Fas-positive peripheral blood lymphocytes.[103-105] Furthermore, apoptosis of peripheral blood lymphocytes is enhanced in vitro by the presence of *P. falciparum*-schizont extract.[100,106] Together, these observations suggest that the lymphocytopenia accompanying malaria infection results from Fas/FasL-mediated apoptosis of T-cells. However, the significance of these findings has been questioned,[107,108] and it is currently uncertain whether T-cell apoptosis during malaria infection is a cause or consequence of disease severity and whether it reflects parasite manipulation or normal host homeostatic down-regulation of lymphocyte clonal expansion following hyper-activation of the immune system. Other studies using the rodent malaria parasite *P. chabaudi* have also reported Fas/FasL-mediated apoptosis of lymphocytes, again primarily T-cells, within the spleens of infected mice.[109,110] Similarly, the functional significance of these observations remains to be determined.

A series of elegant studies using various rodent malaria species has shown that protective *Plasmodium*-specific splenic effector and helper CD4+ T-cells, but not those specific to irrelevant antigens, are deleted following malaria infection.[111-113] Similar investigations have demonstrated that vaccine-induced parasite-specific memory B cells and long-lived plasma cells also undergo apoptosis during malaria infection, although in the latter instance both specific and nonspecific plasma cells are deleted.[114] These observations partly explain why adaptive, long-lasting immunity is difficult to acquire against malaria infection, at least in experimental models. Apoptotic deletion of parasite-specific CD4+ T-cells was initially proposed to be a Fas-mediated event requiring TCR-peptide/MHC complex interaction.[112] However, subsequent work has shown that neither Fas nor TNF-R1 is required for apoptosis of parasite-specific CD4+ T-cells and that deletion of these protective immune cells is an IFN-γ-mediated event.[113] Unpublished studies have also suggested that deletion of CD8+ T-cells providing protection against sporozoite infection of hepatocytes may occur within the liver during malaria infection in rodents.[115]

PCD in Mosquito Vector Cells Associated with Malaria Parasite Infection

Unlike malaria infection within the vertebrate host, *Plasmodium* development within the mosquito vector is not characterized by prolonged periods of obligate intracellular development.[8] Accordingly, there is little evidence that malaria parasites actively suppress PCD of the cells of their mosquito vectors. In contrast, there is increasing evidence that malaria infection is associated with the induction of apoptosis in various mosquito vector cells. However, whether apoptosis of mosquito vector cells is adaptive for the malaria parasite, an incidental by-product of infection, or part of a defensive response by the mosquito vector has yet to be determined.

Ookinete Invasion of the Mosquito Midgut Epithelium

Several studies using different parasite-vector combinations, including the human malaria *P. falciparum*, have demonstrated that ookinete-invaded mosquito midgut epithelial cells undergo apoptosis, exhibiting condensed and TUNEL-positive nuclei, caspase-3 activation and a breakdown of membrane impermeability (as determined by the up-take of cell-impermeant dyes such as ethidium homodimer, YO-PRO-1 and propidium iodide).[116-119] Transcriptomic studies have also identified a number of mosquito apoptosis-related genes whose expression is differentially regulated following ookinete invasion of the midgut epithelium, including several caspases, a death domain-containing protein, a Bax-like inhibitor of apoptosis and several inhibitor-of-apoptosis proteins.[120-122] Abraham et al[120] have shown that the transcriptionally up-regulated caspase-7 of *A. gambiae* is also proteolytically activated at the time of *P. berghei*-ookinete invasion of the midgut epithelium. In some malaria parasite-mosquito vector combinations, ookinete-invaded midgut epithelial cells undergoing apoptosis exhibit elevated expression of nitric oxide synthase, followed by a two-step nitration reaction mediated by inducible peroxidases.[117,123,124] However, nitration of ookinete-invaded midgut epithelial cells is not observed in other malaria parasite-mosquito vector combinations.[123] A mosquito serine protease inhibitor, SRPN10, is also known to undergo nuclear translocation and over-expression in *P. berghei*-invaded midgut epithelial cells undergoing apoptosis, although the function of this molecule is currently unknown.[125] It is also currently uncertain whether all ookinete-invaded midgut epithelial cells undergo apoptosis, or whether a minority of midgut epithelial cells may die through necrosis.[116,119,123] Unlike sporozoite invasion of hepatocytes and merozoite invasion of erythrocytes, ookinete invasion of midgut epithelial cells apparently proceeds without the formation of a parasitophorous vacuole.[8,126] This distinction presumably reflects differences in the strategy of the parasite during these different developmental stages (i.e., intra- versus extracellular development) and the requirement (or lack thereof) for the malaria parasite to maintain a viable host cell.

However, the functional significance of the apoptosis of ookinete-invaded midgut epithelial cells is not yet understood: It may be an attempted defense against the invading ookinete or an act of deliberate induction by ookinetes that facilitates the parasitic infection.[127] Alternatively, apoptosis of ookinete-invaded midgut epithelial cells may represent a general mechanism of midgut epithelial repair, rather than a parasite-specific response, through which damaged cells are removed from the midgut epithelium.[116] In any case, the extrusion of apoptotic midgut epithelial cells into the midgut lumen requires the ookinete to follow both intra- and intercellular routes during its migration across the midgut epithelium and it may account for the lateral movement of the malaria parasite between multiple, adjacent midgut epithelial cells.[118,128] Furthermore, aggregation and redistribution of host cell actin during extrusion of apoptotic midgut epithelial cells constricts and deforms migrating ookinetes and may prevent their subsequent passage through the midgut epithelium.[116-118]

Plasmodium-Associated Reduction in the Fecundity of the Mosquito Vector

A number of studies in the field and the laboratory have shown that malaria-infected mosquitoes have reduced fecundity, producing fewer and less viable eggs (reviewed in ref. 129). The functional significance of this reduction in fecundity is not yet understood,[130] but in mechanistic terms it results at least in part from the resorption of oocytes developing within the ovaries.[129] Hopwood et al[131] have further shown that oocyte resorption is associated with elevated levels of apoptosis in the follicular epithelium surrounding developing oocytes. In *A. stephensi* mosquitoes infected with the rodent malaria species *P. yoelii*, follicular epithelial cells exhibit significantly higher levels of nuclear condensation and TUNEL positivity.[131] Furthermore, follicular epithelial cells express elevated levels of caspase activation in malaria-infected mosquitoes,[132] while injection of the caspase inhibitor Z-VAD.fmk significantly reduces levels of oocyte resorption.[131] In *A. gambiae*, artificial immune stimulation via injection of LPS or Sephadex beads also induces an increase in apoptosis in follicular epithelial cells, suggesting that immune responses to different classes of pathogen result in reduced egg production.[132]

Conclusions

Rapid progress has been made in the characterization of PCD in the cells of the vertebrate host and mosquito vectors during *Plasmodium* infection, with some potentially profound implications for the development of an effective malaria vaccine. In contrast, the existence of PCD within the malaria parasites themselves remains controversial. However, as is true of other unicellular organisms, an appreciation for the possible occurrence and importance of PCD within the malaria parasites themselves, together with a serious search for evidence of PCD and its possible functions, has only recently begun and there may be many exciting and surprising discoveries ahead.

References

1. Garnham PCC. Malaria Parasites and other Haemosporidia. Oxford: Blackwell Scientific Publications; 1966.
2. Hay SI, Guerra CA, Tatem AJ et al. The global distribution and population at risk of malaria: past, present and future. Lancet Infect Dis 2004; 4:327-336.
3. Snow RW, Guerra CA, Noor AM et al. The global distribution of clinical episodes of Plasmodium falciparum malaria. Nature 2005; 434:214-217.
4. Greenwood BM, Bojang K, Whitty CJ et al. Malaria. Lancet 2005; 365:1487-1498.
5. Hyde JE. Drug-resistant malaria. Trends Parasitol 2005; 21:494-498.
6. Hemingway J, Hawkes NJ, McCarroll L et al. The molecular basis of insecticide resistance in mosquitoes. Insect Biochem Mol Biol 2004; 34:653-665.
7. Bannister L, Mitchell G. The ins, outs and roundabouts of malaria. Trends Parasitol 2003; 19:209-213. See http://archive.bmn.com/supp/part/bannister.html for computer animation.
8. Baton LA, Ranford-Cartwright LC. Spreading the seeds of million-murdering death: metamorphoses of malaria in the mosquito. Trends Parasitol 2005; 21:573-580. See Appendix—Supplementary data at http://dx.doi.org/10.1016/j.pt.2005.09.012 to download computer animation.
9. Frevert U. Sneaking in through the back entrance: the biology of malaria liver stages. Trends Parasitol 2004; 20:417-424. See http://archive.bmn.com/supp/part/frevert.html for computer animation.
10. Frischknecht F, Baldacci P, Martin B et al. Imaging movement of malaria parasites during transmission by Anopheles mosquitoes. Cell Microbiol 2004; 6:687-694.
11. Vanderberg JP, Frevert U. Intravital microscopy demonstrating antibody-mediated immobilisation of Plasmodium berghei sporozoites injected into skin by mosquitoes. Int J Parasitol 2004; 34:991-996.
12. Amino R, Thiberge S, Martin B et al. Quantitative imaging of Plasmodium transmission from mosquito to mammal. Nat Med 2006; 12:220-224.
13. Frevert U, Engelmann S, Zougbede S et al. Intravital observation of Plasmodium berghei sporozoite infection of the liver. PLoS Biol 2005; 3:e.192.
14 Ameisen JC. The origin of programmed cell death. Science 1996; 272:1278-1279.
15. Ameisen JC. On the origin, evolution and nature of programmed cell death: a timeline of four billion years. Cell Death Differ 2002; 9:367-393.
16. Ameisen JC. Looking for death at the core of life in the light of evolution. Cell Death Differ 2004; 11:4-10.
17. Broker LE, Kruyt FA, Giaccone G. Cell death independent of caspases: a review. Clin Cancer Res 2005; 11:3155-3162.
18. Meier P, Finch A, Evan G. Apoptosis in development. Nature 2000; 407:796-801.
19. Vaux DL, Haecker G, Strasser A. An evolutionary perspective on apoptosis. Cell 1994; 76:777-779.
20. Williams GT. Programmed cell death: a fundamental protective response to pathogens. Trends Microbiol 1994; 2:463-464.
21. James ER, Green DR. Manipulation of apoptosis in the host-parasite interaction. Trends Parasitol 2004; 20:280-287.
22. Hamilton WD. Narrow Roads of Gene Land, Volume 1: Evolution of Social Behavior. Oxford: Oxford University Press; 1996.
23. Babiker HA, Ranford-Cartwright LC, Currie D et al. Random mating in a natural population of the malaria parasite Plasmodium falciparum. Parasitology 1994; 109:413-421.
24. Babiker HA, Charlwood JD, Smith T et al. Gene flow and cross-mating in Plasmodium falciparum in households in a Tanzanian village. Parasitology 1995; 111:433-442.
25. Walliker D. Malaria parasites: randomly interbreeding or clonal populations? Parasitol Today 1991; 7:232-235.
26. Anderson TJ, Haubold B, Williams JT et al. Microsatellite markers reveal a spectrum of population structures in the malaria parasite Plasmodium falciparum. Mol Biol Evol 2000; 17:1467-1482.

27. Babiker HA, Walliker D. Current views on the population structure of Plasmodium falciparum: Implications for control. Parasitol Today 1997; 13:262-267.
28. Paul RE, Packer MJ, Walmsley M et al. Mating patterns in malaria parasite populations of Papua New Guinea. Science 1995; 269:1709-1711.
29. Razakandrainibe FG, Durand P, Koella JC et al. Clonal population structure of the malaria agent Plasmodium falciparum in high-infection regions. Proc Natl Acad Sci USA 2005; 102:17388-17393.
30. Walliker D, Babiker H, Ranford-Cartwright L. The genetic structure of malaria parasite populations. In: Sherman IW, ed. Malaria: Parasite Biology, Pathogenesis and Protection. Washington, DC: American Society for Microbiology Press; 1998:235-52.
31. Dawkins R. The Selfish Gene. 2nd ed. Oxford: Oxford University Press; 1989.
32. Dieckmann U, Metz JAJ, Sabelis MW et al. eds. Adaptive Dynamics of Infectious Diseases: In Pursuit of Virulence Management. Cambridge: Cambridge University Press; 2002.
33. Ewald PW. Evolution of Infectious Disease. New York: Oxford University Press; 1994.
34. Sober E, Wilson DS. Unto Others: The Evolution and Psychology of Unselfish Behavior. Cambridge: Harvard University Press; 1998.
35. Deponte M, Becker K. Plasmodium falciparum—do killers commit suicide? Trends Parasitol 2004; 20:165-169.
36. Paul REL, Ariey F, Robert V. The evolutionary ecology of Plasmodium. Ecol Lett 2003; 6:866-880.
37. Dyer M, Day KP. Regulation of the rate of asexual growth and commitment to sexual development by diffusible factors from in vitro cultures of Plasmodium falciparum. Am J Trop Med Hyg 2003; 68:403-409.
38. Hurd H, Carter V, Nacer A. Interactions between malaria and mosquitoes: the role of apoptosis in parasite establishment and vector response to infection. Curr Top Microbiol Immunol 2005; 289:185-217.
39. Hurd H, Carter V. The role of programmed cell death in Plasmodium-mosquito interactions. Int J Parasitol 2004; 34:1459-1472.
40. Uren AG, O'Rourke K, Aravind LA et al. Identification of paracaspases and metacaspases: two ancient families of caspase-like proteins, one of which plays a key role in MALT lymphoma. Mol Cell 2000; 6:961-967.
41. van de Sand C, Horstmann S, Schmidt A et al. The liver stage of Plasmodium berghei inhibits host cell apoptosis. Mol Microbiol 2005; 58:731-742.
42. Wu Y, Wang X, Liu X et al. Data-mining approaches reveal hidden families of proteases in the genome of malaria parasite. Genome Res 2003; 13:601-616.
43. Sperandio S, Poksay K, de BI et al. Paraptosis: mediation by MAP kinases and inhibition by AIP-1/Alix. Cell Death Differ 2004; 11:1066-1075.
44. Ward P, Equinet L, Packer J et al. Protein kinases of the human malaria parasite Plasmodium falciparum: the kinome of a divergent eukaryote. BMC Genomics 2004; 5:79.
45. Picot S, Burnod J, Bracchi V et al. Apoptosis related to chloroquine sensitivity of the human malaria parasite Plasmodium falciparum. Trans R Soc Trop Med Hyg 1997; 91:590-591.
46. Nyakeriga AM, Perlmann H, Hagstedt M et al. Drug-induced death of the asexual blood stages of Plasmodium falciparum occurs without typical signs of apoptosis. Microbes Infect 2006; in press.
47. Pankova-Kholmyansky I, Dagan A, Gold D et al. Ceramide mediates growth inhibition of the Plasmodium falciparum parasite. Cell Mol Life Sci 2003; 60:577-587.
48. Pettus BJ, Chalfant CE, Hannun YA. Ceramide in apoptosis: an overview and current perspectives. Biochim Biophys Acta 2002; 1585:114-125.
49. Al-Olayan EM, Williams GT, Hurd H. Apoptosis in the malaria protozoan, Plasmodium berghei: a possible mechanism for limiting intensity of infection in the mosquito. Int J Parasitol 2002; 32:1133-1143.
50. Al-Olayan EM, Williams GT, Hurd H. Erratum to Apoptosis in the malaria protozoan, Plasmodium berghei: a possible mechanism for limiting intensity of infection in the mosquito [Int J Parasitol. 32(9) (2002) 1133-1143]. Int J Parasitol 2003; 33:105.
51. Leirião P, Albuquerque SS, Corso S et al. HGF/MET signalling protects Plasmodium-infected host cells from apoptosis. Cell Microbiol 2005; 7:603-609.
52. Carrolo M, Giordano S, Cabrita-Santos L et al. Hepatocyte growth factor and its receptor are required for malaria infection. Nat Med 2003; 9:1363-1369.
53. Leiriao P, Mota MM, Rodriguez A. Apoptotic Plasmodium-infected hepatocytes provide antigens to liver dendritic cells. J Infect Dis 2005; 191:1576-1581.
54. James E. Apoptosis: key to the attenuated malaria vaccine? J Infect Dis 2005; 191:1573-1575.
55. Leiriao P, Mota MM, Rodriguez A. Do apoptotic Plasmodium-infected hepatocytes initiate protective immune responses? Reply to Rénia et al. J Infect Dis 2006; 193:164-165.
56. Rénia L, Maranon C, Hosmalin A et al. Do apoptotic Plasmodium-infected hepatocytes initiate protective immune responses? J Infect Dis 2006; 193:163-164.

57. van Dijk MR, Douradinha B, Franke-Fayard B et al. Genetically attenuated, P36p-deficient malaria sporozoites induce protective immunity and apoptosis of infected liver cells. Proc Natl Acad Sci USA 2005; 102:12194-12199.
58. Heussler VT, Kuenzi P, Rottenberg S. Inhibition of apoptosis by intracellular protozoan parasites. Int J Parasitol 2001; 31:1166-1176.
59. Luder CG, Gross U, Lopes MF. Intracellular protozoan parasites and apoptosis: diverse strategies to modulate parasite-host interactions. Trends Parasitol 2001; 17:480-486.
60. Daugas E, Cande C, Kroemer G. Erythrocytes: death of a mummy. Cell Death Differ 2001; 8:1131-1133.
61. Lang KS, Lang PA, Bauer C et al. Mechanisms of suicidal erythrocyte death. Cell Physiol Biochem 2005; 15:195-202.
62. Berg CP, Engels IH, Rothbart A et al. Human mature red blood cells express caspase-3 and caspase-8, but are devoid of mitochondrial regulators of apoptosis. Cell Death Differ 2001; 8:1197-1206.
63. Bratosin D, Estaquier J, Petit F et al. Programmed cell death in mature erythrocytes: a model for investigating death effector pathways operating in the absence of mitochondria. Cell Death Differ 2001; 8:1143-1156.
64. Lang F, Lang KS, Wieder T et al. Cation channels, cell volume and the death of an erythrocyte. Pflugers Arch 2003; 447:121-125.
65. Lang KS, Duranton C, Poehlmann H et al. Cation channels trigger apoptotic death of erythrocytes. Cell Death Differ 2003; 10:249-256.
66. Bratosin D, Mazurier J, Tissier JP et al. Cellular and molecular mechanisms of senescent erythrocyte phagocytosis by macrophages. A review Biochimie 1998; 80:173-195.
67. Boas FE, Forman L, Beutler E. Phosphatidylserine exposure and red cell viability in red cell aging and in hemolytic anemia. Proc Natl Acad Sci USA 1998; 95:3077-3081.
68. Lang F, Lang PA, Lang KS et al. Channel-induced apoptosis of infected host cells—the case of malaria. Pflugers Arch 2004; 448:319-324.
69. Brand VB, Sandu CD, Duranton C et al. Dependence of Plasmodium falciparum in vitro growth on the cation permeability of the human host erythrocyte. Cell Physiol Biochem 2003; 13:347-356.
70. Staines HM, Chang W, Ellory JC et al. Passive Ca2+ transport and Ca2+-dependent K+ transport in Plasmodium falciparum-infected red cells. J Membr Biol 1999; 172:13-24.
71. Weil M, Jacobson MD, Raff MC. Are caspases involved in the death of cells with a transcriptionally inactive nucleus? Sperm and chicken erythrocytes. J Cell Sci 1998; 111:2707-2715.
72. Berendt AR, Ferguson DJ, Newbold CI. Sequestration in Plasmodium falciparum malaria: sticky cells and sticky problems. Parasitol Today 1990; 6:247-254.
73. Newbold CI. Antigenic variation in Plasmodium falciparum: mechanisms and consequences. Curr Opin Microbiol 1999; 2:420-425.
74. Eda S, Sherman IW. Cytoadherence of malaria-infected red blood cells involves exposure of phosphatidylserine. Cell Physiol Biochem 2002; 12:373-384.
75. Lang KS, Roll B, Myssina S et al. Enhanced erythrocyte apoptosis in sickle cell anemia, thalassemia and glucose-6-phosphate dehydrogenase deficiency. Cell Physiol Biochem 2002; 12:365-372.
76. Beeson JG, Duffy PE. The immunology and pathogenesis of malaria during pregnancy. Curr Top Microbiol Immunol 2005; 297:187-227.
77. Grobusch MP, Kremsner PG. Uncomplicated malaria. Curr Top Microbiol Immunol 2005; 295:83-104.
78. Hunt NH, Golenser J, Chan-Ling T et al. Immunopathogenesis of cerebral malaria. Int J Parasitol 2006; 36:569-582.
79. Idro R, Jenkins NE, Newton CR. Pathogenesis, clinical features and neurological outcome of cerebral malaria. Lancet Neurol 2005; 4:827-840.
80. Planche T, Dzeing A, Ngou-Milama E et al. Metabolic complications of severe malaria. Curr Top Microbiol Immunol 2005; 295:105-136.
81. Roberts DJ, Casals-Pascual C, Weatherall DJ. The clinical and pathophysiological features of malaria anaemia. Curr Top Microbiol Immunol 2005; 295:137-167.
82. Schofield L, Grau GE. Immunological processes in malaria pathogenesis. Nat Rev Immunol 2005; 5:722-735.
83. Sherman IW, Eda S, Winograd E. Cytoadherence and sequestration in Plasmodium falciparum: defining the ties that bind. Microbes Infect 2003; 5:897-909.
84. Pino P, Taoufiq Z, Nitcheu J et al. Blood-brain barrier breakdown during cerebral malaria: suicide or murder? Thromb Haemost 2005; 94:336-340.
85. Hemmer CJ, Lehr HA, Westphal K et al. Plasmodium falciparum malaria: reduction of endothelial cell apoptosis in vitro. Infect Immun 2005; 73:1764-1770.
86. Pino P, Vouldoukis I, Dugas N et al. Redox-dependent apoptosis in human endothelial cells after adhesion of Plasmodium falciparum-infected erythrocytes. Ann N Y Acad Sci 2003; 1010:582-586.

87. Pino P, Vouldoukis I, Kolb JP et al. Plasmodium falciparum-infected erythrocyte adhesion induces caspase activation and apoptosis in human endothelial cells. J Infect Dis 2003; 187:1283-1290.
88. Potter S, Chan-Ling T, Ball HJ et al. Perforin mediated apoptosis of cerebral microvascular endothelial cells during experimental cerebral malaria. Int J Parasitol 2006; 36:485-496.
89. Wassmer SC, Combes V, Candal FJ et al. Platelets potentiate brain endothelial alterations induced by Plasmodium falciparum. Infect Immun 2006; 74:645-653.
90. Wiese L, Kurtzhals JA, Penkowa M. Neuronal apoptosis, metallothionein expression and proinflammatory responses during cerebral malaria in mice. Exp Neurol 2006.
91. Wassmer SC, de Souza JB, Frere C et al. TGF-β1 released from activated platelets can induce TNF-stimulated human brain endothelium apoptosis: a new mechanism for microvascular lesion during cerebral malaria. J Immunol 2006; 176:1180-1184.
92. Schluesener HJ, Kremsner PG, Meyermann R. Widespread expression of MRP8 and MRP14 in human cerebral malaria by microglial cells. Acta Neuropathol (Berl) 1998; 96:575-580.
93. Kaiser K, Texier A, Ferrandiz J et al. Recombinant human erythropoietin prevents the death of mice during cerebral malaria. J Infect Dis 2006; 193:987-995.
94. Medana IM, Mai NT, Day NP et al. Cellular stress and injury responses in the brains of adult Vietnamese patients with fatal Plasmodium falciparum malaria. Neuropathol Appl Neurobiol 2001; 27:421-433.
95. Potter SM, Chan-Ling T, Rosinova E et al. A role for Fas-Fas ligand interactions during the late-stage neuropathological processes of experimental cerebral malaria. J Neuroimmunol 2006; 173:96-107.
96. Crocker IP, Tanner OM, Myers JE et al. Syncytiotrophoblast degradation and the pathophysiology of the malaria-infected placenta. Placenta 2004; 25:273-282.
97. Guha M, Kumar S, Choubey V et al. Apoptosis in liver during malaria: role of oxidative stress and implication of mitochondrial pathway. FASEB J 2006.
98. Achtman AH, Bull PC, Stephens R et al. Longevity of the immune response and memory to blood-stage malaria infection. Curr Top Microbiol Immunol 2005; 297:71-102.
99. Balde AT, Sarthou JL, Roussilhon C. Acute Plasmodium falciparum infection is associated with increased percentages of apoptotic cells. Immunol Lett 1995; 46:59-62.
100. Balde AT, Aribot G, Tall A et al. Apoptosis modulation in mononuclear cells recovered from individuals exposed to Plasmodium falciparum infection. Parasite Immunol 2000; 22:307-318.
101. Kemp K, Akanmori BD, Adabayeri V et al. Cytokine production and apoptosis among T-cells from patients under treatment for Plasmodium falciparum malaria. Clin Exp Immunol 2002; 127:151-157.
102. Kemp K, Akanmori BD, Kurtzhals JA et al. Acute P. falciparum malaria induces a loss of CD28- T IFN-gamma producing cells. Parasite Immunol 2002; 24:545-548.
103. Kern P, Dietrich M, Hemmer C et al. Increased levels of soluble Fas ligand in serum in Plasmodium falciparum malaria. Infect Immun 2000; 68:3061-3063.
104. Matsumoto J, Kawai S, Terao K et al. Malaria infection induces rapid elevation of the soluble Fas ligand level in serum and subsequent T lymphocytopenia: possible factors responsible for the differences in susceptibility of two species of Macaca monkeys to Plasmodium coatneyi infection. Infect Immun 2000; 68:1183-1188.
105. Riccio EK, Junior IN, Riccio LR et al. Malaria associated apoptosis is not significantly correlated with either parasitemia or the number of previous malaria attacks. Parasitol Res 2003; 90:9-18.
106. ToureBalde A, Sarthou JL, Aribot G et al. Plasmodium falciparum induces apoptosis in human mononuclear cells. Infect Immun 1996; 64:744-750.
107. Hviid L, Kemp K. What is the cause of lymphopenia in malaria? Infect Immun 2000; 68:6087-6089.
108. Kemp K, Akanmori BD, Hviid L. West African donors have high percentages of activated cytokine producing T-cells that are prone to apoptosis. Clin Exp Immunol 2001; 126:69-75.
109. Helmby H, Jonsson G, Troye-Blomberg M. Cellular changes and apoptosis in the spleens and peripheral blood of mice infected with blood-stage Plasmodium chabaudi chabaudi AS. Infect Immun 2000; 68:1485-1490.
110. Sanchez-Torres L, Rodriguez-Ropon A, Aguilar-Medina M et al. Mouse splenic CD4+ and CD8+ T-cells undergo extensive apoptosis during a Plasmodium chabaudi chabaudi AS infection. Parasite Immunol 2001; 23:617-626.
111. Hirunpetcharat C, Good MF. Deletion of Plasmodium berghei-specific CD4+ T-cells adoptively transferred into recipient mice after challenge with homologous parasite. Proc Natl Acad Sci USA 1998; 95:1715-1720.
112. Wipasa J, Xu H, Stowers A et al. Apoptotic deletion of Th cells specific for the 19-kDa carboxyl-terminal fragment of merozoite surface protein 1 during malaria infection. J Immunol 2001; 167:3903-3909.
113. Xu H, Wipasa J, Yan H et al. The mechanism and significance of deletion of parasite-specific CD4(+) T-cells in malaria infection. J Exp Med 2002; 195:881-892.
114. Wykes MN, Zhou YH, Liu XQ et al. Plasmodium yoelii can ablate vaccine-induced long-term protection in mice. J Immunol 2005; 175:2510-2516.

115. Krzych U, Schwenk J. The dissection of CD8 T-cells during liver-stage infection. Curr Top Microbiol Immunol 2005; 297:1-24.
116. Baton LA, Ranford-Cartwright LC. Plasmodium falciparum ookinete invasion of the midgut epithelium of Anopheles stephensi is consistent with the Time Bomb model. Parasitology 2004; 129:663-676.
117. Han YS, Thompson J, Kafatos FC et al. Molecular interactions between Anopheles stephensi midgut cells and Plasmodium berghei: the time bomb theory of ookinete invasion of mosquitoes. EMBO J 2000; 19:6030-6040.
118. Vlachou D, Zimmermann T, Cantera R et al. Real-time, in vivo analysis of malaria ookinete locomotion and mosquito midgut invasion. Cell Microbiol 2004; 6:671-685.
119. Zieler H, Dvorak JA. Invasion in vitro of mosquito midgut cells by the malaria parasite proceeds by a conserved mechanism and results in death of the invaded midgut cells. Proc Natl Acad Sci USA 2000; 97:11516-11521.
120. Abraham EG, Islam S, Srinivasan P et al. Analysis of the Plasmodium and Anopheles transcriptional repertoire during ookinete development and midgut invasion. J Biol Chem 2004; 279:5573-5580.
121. Vlachou D, Schlegelmilch T, Christophides GK et al. Functional genomic analysis of midgut epithelial responses in Anopheles during Plasmodium invasion. Curr Biol 2005; 15:1185-1195.
122. Xu X, Dong Y, Abraham EG et al. Transcriptome analysis of Anopheles stephensi-Plasmodium berghei interactions. Mol Biochem Parasitol 2005; 142:76-87.
123. Gupta L, Kumar S, Han YS et al. Midgut epithelial responses of different mosquito-Plasmodium combinations. The actin cone zipper repair mechanism in Aedes aegypti. Proc Natl Acad Sci USA 2005; 102:4010-4015.
124. Kumar S, Gupta L, Han YS et al. Inducible peroxidases mediate nitration of anopheles midgut cells undergoing apoptosis in response to Plasmodium invasion. J Biol Chem 2004; 279:53475-53482.
125. Danielli A, Barillas-Mury C, Kumar S et al. Overexpression and altered nucleocytoplasmic distribution of Anopheles ovalbumin-like SRPN10 serpins in Plasmodium-infected midgut cells. Cell Microbiol 2005; 7:181-190.
126. Yuda M, Ishino T. Liver invasion by malaria parasites—how do malaria parasites break through the host barrier? Cell Microbiol 2004; 6:1119-1125.
127. Han YS, Barillas-Mury C. Implications of Time Bomb model of ookinete invasion of midgut cells. Insect Biochem Mol Biol 2002; 32:1311-1316.
128. Baton LA, Ranford-Cartwright LC. How do malaria ookinetes cross the mosquito midgut wall? Trends Parasitol 2005; 21:22-28.
129. Hurd H. Manipulation of medically important insect vectors by their parasites. Annu Rev Entomol 2003; 48:141-161.
130. Hurd H. Host fecundity reduction: a strategy for damage limitation? Trends Parasitol 2001; 17:363-368.
131. Hopwood JA, Ahmed AM, Polwart A et al. Malaria-induced apoptosis in mosquito ovaries: a mechanism to control vector egg production. J Exp Biol 2001; 204:2773-2780.
132. Ahmed AM, Hurd H. Immune stimulation and malaria infection impose reproductive costs in Anopheles gambiae via follicular apoptosis. Microbes Infect 2006; 8:308-315.

CHAPTER 8

In Search of Atropos' Scissors:
Severing the Life-Thread of *Plasmodium*

Marcel Deponte*

Abstract

Protozoa of the genus *Plasmodium* are interesting models to study the mode of cell death of unicellular organisms. It is well known that the malaria-causing parasites can be killed in vitro and that they also die in vivo. The central question is how does cell death occur in *Plasmodium*? To date, the hypothesis that some stages of malaria parasites are able to undergo a form of programmed cell death is supported by available data, but there is no evidence showing that certain proteins are required for the observed processes. Here we present the current knowledge on *Plasmodium* metacaspases, because these putative proteases are the most promising candidates that might be essential for the execution of a programmed cell death in malaria parasites.

Introduction

In Greek mythology the goddess Atropos, one of the three sisters of fate, chose the mechanism of death and ended the life of each mortal by cutting the thread of life when the proper time has come for death. In case of most unicellular organisms it is unknown how cell death occurs, or in other words what kind of scissors Atropos uses to sever their life-thread. Are malaria-causing parasites able to undergo a kind of programmed cell death? The answer of this question is closely related to the identification and characterization of genes and proteins that are required for the execution of cell death. So far apoptotic markers including chromatin condensation, DNA fragmentation, externalization of phosphatidylserine and decreased glutathione concentrations as well as morphological changes such as cell shrinkage, membrane blebbing and nuclear fragmentation have been detected in different *Plasmodium* spp. supporting the hypothesis, that malaria parasites are able to undergo a form of programmed cell death (for review see refs. 1 and 2). However, in contrast to the yeast *Saccharomyces cerevisiae*[3,4] no genetic methods have been used so far to proof the existence of a molecular machinery that is involved in the controlled cell death of *Plasmodium*. In *S. cerevisiae* the metacaspase Yca1 has been shown to play a central role as an executor of yeast apoptosis.[4] Metacaspases in general are (putative) proteases that have been identified in *silico*[5] and that are related to caspases.[6] Here we present the current knowledge on the *Plasmodium* metacaspases and discuss their potential as executors of cell death.

Things to Know about *Plasmodium* and Malaria

In contrast to the phylogenetically distant trypanosomatids, *Plasmodium* spp. belong together with *Toxoplasma* and *Cryptosporidium* to the apicomplexan group of parasites. Malaria parasites infect reptiles, birds or mammals (including men, monkeys or rodents) and are transferred from one intermediate host to the other by mosquitoes of the genera *Anopheles*, *Aedes* and *Culex*. The mosquitoes are regarded as the definitive host because the formation of the diploid zygote and

*Marcel Deponte—Adolf Butenandt-Institute for Physiological Chemistry, Ludwig Maximilians University, Butenandtstr. 5, D-81377 Munich, Germany. Email: marcel.deponte@gmx.de

Programmed Cell Death in Protozoa, edited by José Manuel Pérez Martín.
©2008 Landes Bioscience and Springer Science+Business Media.

following meiosis occur in their midgut. All other *Plasmodium* parasite stages are haploid and replicate asexually in the mosquito (sporogony) or in the vertebrate (schizogony) (for an overview on *Plasmodium* biology see ref. 7).

Four species of *Plasmodium* infect men, of which three (*P. vivax, P. ovale, P. malariae*) are rarely fatal, whereas *P. falciparum* is much more pathogenic: In 2002 the causative agent of tropical malaria was estimated to be responsible for 515 (range 300-660) million clinical cases, especially in Africa (70%) and South East Asia (25%).[8] Even though only a small proportion of these infections lead to severe complications, it is believed that at least one million deaths occur from malaria each year. The vast majority of deaths caused by *P. falciparum* occur among young children in sub-Saharan Africa.[9]

Presently there is no vaccine available against malaria but a variety of different chemical compounds such as antifolates, artemisinin derivatives and amino quinolines have been highly effective in curing the disease.[10] Furthermore, the insecticide dichlorodiphenyl-trichloroethane (DDT) efficiently kills *Anopheles* mosquitoes and interrupts disease transmission. The usage of DDT is actually one of the most important reasons why malaria was almost eradicated in many parts of the world throughout the 1960s.[11] Unfortunately, this plague is on the rise again due to resistance of malaria parasites and mosquitoes against antimalarials and insecticides respectively, in addition to economic and social problems (such as debates leading to a stop of DDT spraying in houses). To learn as much as possible about the biology and the biochemistry of *Plasmodium* and *Anopheles* is therefore not only interesting from a scientific point of view but might also help to ameliorate this situation in some ways.

The latest investigations on *Plasmodium* species benefit particularly from four genome sequencing projects that have already been published, including *P. falciparum* (strain 3D7)[12] and the rodent malaria parasites *P. yoelii* (strain 17XNL),[13] *P. berghei* and *P. chabaudi* (strains ANKA and AS, respectively).[14] These data, in addition to data from ongoing sequencing projects for *P. knowlesi, P. gallinaceum* and *P. reichenowi* are available online (PlasmoDB@http://plasmodb. org/,[15] http://www.genedb.org/, http://www.sanger.ac.uk/Projects/Protozoa/ and http://www. tigr.org/tdb/e2k1/pya1/) and are widely used by the scientific community. The genetic manipulation of different *Plasmodium* spp. has been established. For example, in the case of *P. falciparum* the asexual blood stages can be manipulated,[16] and for *P. berghei* it is even possible to analyze the whole life cycle.[17]

Metacaspases as Putative Executors of Cell Death

Metacaspases occur in unicellular organisms and plants.[5] In analogy to metazoan caspases, metacaspases are likely to belong to the C14 family of clan CD enzymes possessing a conserved (putative) caspase-like catalytic dyad that is composed of a histidine and a cysteine residue.[6] Caspases play an essential role during the programmed cell death of metazoa and are specific for aspartate as the C-terminal amino acid (P1) of a tetrapeptide motif.[18] Is this also true for metacaspases?

So far no proteolytic activity of a recombinant or purified metacaspase from a unicellular organism has been reported, but studies on metacaspases Atmc4, Atmc9[19] and Atmcp1b, Atmcp2b[20] from the plant *Arabidopsis thaliana* suggest an arginine/lysine specific proteolytic activity. Enzymatic activity and autocatalytic processing of Atmc4 and Atmc9 depend on Cys_{139} and Cys_{147} respectively, being part of the predicted catalytic dyad. In contrast to canonical caspases, Atmc4 and Atmc9 do not cleave the aspartate-containing peptides DEVD, YVAD and VAD.[19] The metacaspase mcII-Pa from Norway spruce (*Picea abies*), which is involved in programmed cell death during plant embryogenesis, also prefers basic residues as C-terminal amino acid and does not cleave the aspartate-containing peptides VEID and YVAD. Enzymatic activity and autocatalytic processing of mcII-Pa are dependent on Cys_{139} of the predicted catalytic dyad.[21] Surprisingly, cell death in Norway spruce embryos is coupled to a caspase-like enzymatic activity and both cell death and VEIDase activity are partially inhibited by *mcII-Pa* RNA interference.[22] Hydrogen peroxide- or aging-induced apoptosis in *S. cerevisiae* is also coupled to a caspase-like enzymatic activity and can be abrogated by disruption of the metacaspase gene *YCA1*.[4] Furthermore, caspase activity coupled cell death in *S. cerevisiae* and in the midgut stages of *P. berghei* can be partially inhibited by using aspartate(!)-containing suicide inhibitors such as N-benzyloxy-carbonyl-VAD fluoromethyl ketone (zVAD.fmk).[4,23]

In view of these experimental data—and after having a closer look at alignments of the highly conserved core domain of metacaspases, starting with a basic residue (Fig. 1)—it is

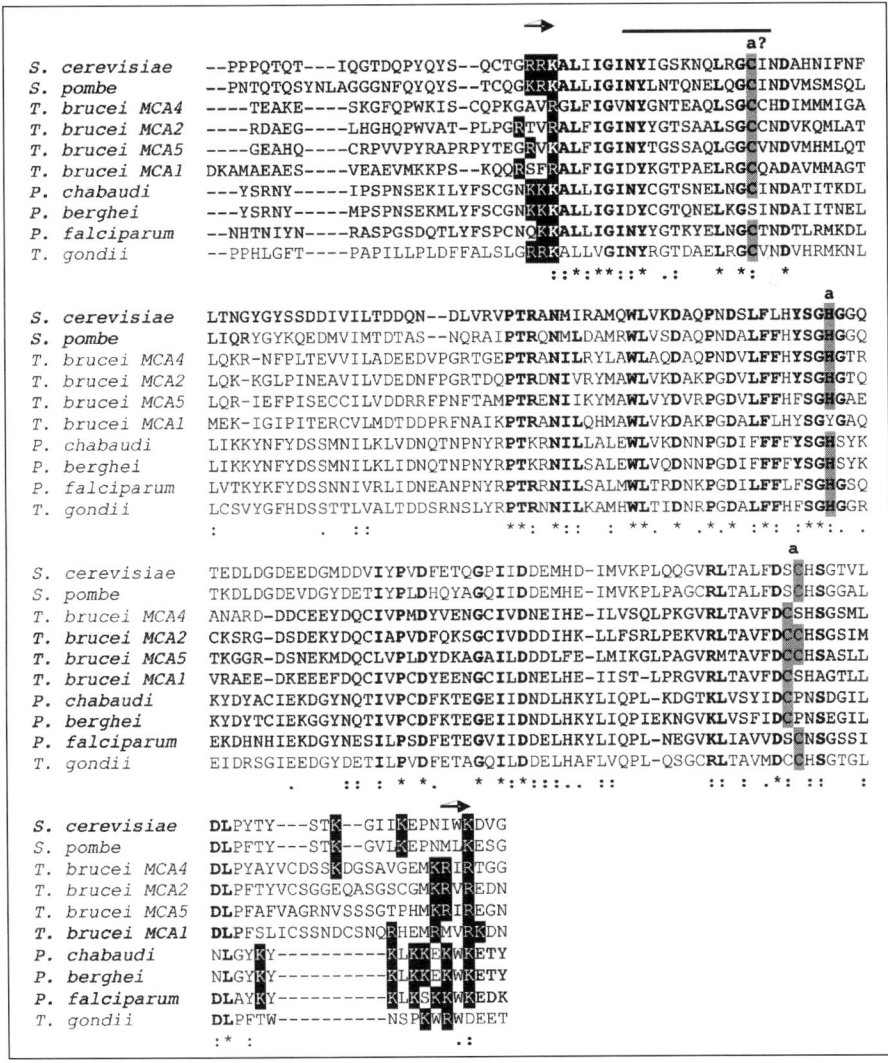

Figure 1. Multiple sequence alignment of the putative large (p20) subunit of metacaspases from unicellular organisms. Patches of basic residues at the N- and C-terminus of the highly conserved core domain (arrows) are shaded in black. Residues forming the putative active site are shaded in grey. Residues forming loop 1 at the active site are marked with a bar. *P. chabaudi*, gi|70947027; *P. berghei*, gi|29788140; *P. yoelii*, (fragment) gi|82594051; *P. falciparum*, gi|85822146; *Toxoplasma gondii*, TgTwinScan_3998 (ToxoDB annotation http://ToxoDB.org); *Trypanosoma brucei* MCA1, gi|19032264; *T. brucei* MCA2, gi|19032266; *T. brucei*, MCA4 gi|19032270; *T. brucei*, MCA5 gi|24475397.

very likely that proteolytically active metacaspases are in general specific for arginine/lysine residues. Therefore, further experiments are required to understand how a metacaspase dependent cell death and a caspase-like aspartate specific protease activity in plants, yeast and maybe *Plasmodium* can be linked.

Metacaspase Structure and Function: The Devil Is in the Details

Another complication with respect to the proteolytic activity of metacaspases arises when looking at the putative active site residues (Fig. 1): For example a tyrosine substitutes for the putative catalytic histidine in the *Trypanosoma brucei* metacaspase MCA1 and in the metacaspase-like proteins from *P. berghei* and *P. falciparum* (annotations gi|29788142 and PF14_0363, respectively). In the case of the cysteine residue of the putative catalytic dyad, the situation is even more diverse leading to three different possibilities:

1. Several metacaspases (for example from yeast, *T. gondii* and *P. falciparum*) have a caspase-like cysteine residue in a D[SCA]*C*[HN]S-motif. In contrast, this cysteine residue is replaced by a serine or proline residue in other metacaspases (for example from *P. berghei* and *P. chabaudi* but also MCA1 and MCA4 from *T. brucei*). Do these proteins possess different functions? Although this possibility cannot be excluded in the case of the *T. brucei* metacaspases, it would be highly surprising as far as the closely related *Plasmodium* metacaspases are concerned, because—in contrast to trypanosomes—malaria parasites seem to possess only one metacaspase gene each. (The *P. falciparum* and *P. berghei* metacaspase-like proteins mentioned above have a DT*TY*[TS]-motif instead of the D[SCA]*C*[HN]S-motif and are not canonical metacaspases.)
2. Alexander Szallies et al pointed out that there is another metacaspase-specific cysteine residue (see G*C*xxD-motif in Fig. 1) that might be located at the active site.[24] On the one hand a molecular model of the *P. falciparum* metacaspase supports this hypothesis, because Cys$_{337}$ is located in loop 1 of the putative active site and is in similar distance to His$_{404}$ when compared to Cys$_{460}$ (Fig. 2). On the other hand, the metacaspase from *P. berghei* for example is lacking a G*C*xxD and a D[SCA]*C*[HN]S-motif.
3. Maybe the cysteine residue adjacent to the serine or proline residue of several metacaspases (Fig. 1) adopts a similar position in the tertiary structure because of its localization at the presumably flexible loop 2 of the putative active site (Fig. 2). This cysteine residue is also present in the D*C*PNS-motif of the metacaspases from *P. berghei* and *P. chabaudi*.

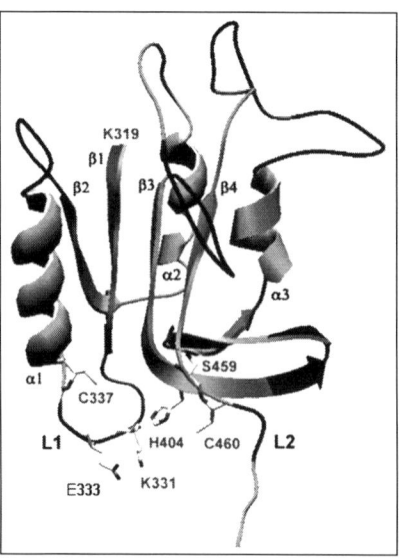

Figure 2. Molecular model of the conserved core of the *P. falciparum* metacaspase (PfMca) based on the X-ray structure of human caspase 3 in complex with the inhibitor XIAP.[26] In contrast to human caspase 3, PfMca possesses several insertions presumably located in loop areas (dark newly structured areas). Substrate binding loops L1 and L2 contain cysteine residues Cys$_{337}$ and Cys$_{460}$, respectively. Distances between His$_{404}$ and both cysteine residues of this (inhibited) model are 6-8 Å. The nomenclature for β-strands was adopted from caspase 3. Computations of the model were carried out at the Swiss-Model server using the optimize (project) mode.[27] The force field energy of the model (−5.8 MJ/mol) was calculated with the GROMOS96 implementation of Swiss-PDB Viewer.

The Metacaspase Core Domain Is Flanked by a *Plasmodium* Specific N-Terminal Sequence

The C- and N-terminal sequences flanking the conserved core domain differ significantly among the metacaspases from plants and unicellular organisms.[5,6] The highlighted basic residue preceding the putative catalytic domain (Fig. 3) is well conserved among metacaspases from unicellular organisms, whereas the first 315 residues of the very large *Plasmodium* metacaspases share no significant similarities with other metacaspases known so far. Furthermore, there is only little sequence similarity between rodent and *P. falciparum* metacaspases as far as residues ~150-305 are concerned. In particular, the N-terminal domain of the *P. falciparum* protein is rich in asparagine residues (which is a typical feature of many *P. falciparum* proteins[25]). Several asparagine-rich *P. falciparum* proteins have an innate predisposition to form self-propagating amyloid fibers.[25] This might facilitate protein-protein interactions and lead to an increased local metacaspase concentration.

A closer look at the N-terminus of the *Plasmodium* metacaspases reveals a pattern that is composed of basic and aromatic residues. A search for the pattern 'KxY[FHWY]xx KK[FHWY][KRH]' in PlasmoDB[15] lead to the identification of four sequences. Two of them are encoded by the adjacent genes PF13_0288 and PF13_0289 on chromosome 13. PF13_0289 encodes the metacaspase (gi|85822146), whereas PF13_0288 encodes a 49 kDa hypothetical protein with unknown function that is also well conserved in *P. chabaudi* (gi|56522539) and *P. berghei* (gi|56500255). Presumably both genes are functionally linked (at least it seems very unlikely that two completely different proteins coincidentally share an amino acid motif and are encoded by two adjacent genes). Future work will tell us whether both genes are related to a programmed cell death in *Plasmodium* spp.

```
P. chabaudi    MAQIYVKVHELKFVNNSERNSHYVKIYWDDKKYKSQTKDGGHYIFNETFLIPITN---IN 57
P. berghei     MDKIYVKVHELKFVNNLERNSHYVKIYWDDKKYKSQTKDGGCYIFNETFLIPITN---IY 57
P. yoelii      MDQIYVKAHELKFVNNLERNSHYVKIYWDDKKYKSQTKDGGCYIFNENFLIPITN---IY 57
P. falciparum  MEKIYVKIYELSGLEDKDNFSCYIKIYWQNKKYKSCILQKNPYKFNEIFLLPIDIKNNVK 60
               * :**** :**. ::: :. * *:****:***** :. * *** **:** :

P. chabaudi    YQNDQIIYVEVWESNLLN-KQCAYTIFTLNNIKTGQIIKENIALIEVLKKCTLELSINIV 116
P. berghei     DQKDQIIYVEIWESNLLN-KQCAYTFFTLNSIKIGQIIKENITFIEVLKKCTLGLSINIV 116
P. yoelii      DQKDQLIYVEIWESNLLN-KQCAYTFFTLNSIKIGQIIKENITFIEVLKKCTLELSININ 116
P. falciparum  DEKNNILSIEVWSSGILNNNKIAYTFFELDHIRRERISSEKINLIDVVKKCTLQISVHII 120
               ::::::: :*:*.*.:** :: ***:* *: :: :* .*:* :*:*:***** :*:*

P. chabaudi    RNQKDILFFNIKESLPTYQ-DQQIRNAIWENE-DEASIIKQLININKFNGITNLGDYKNS 174
P. berghei     RNQKDILFFNIKELLPTFQ-DQEIINAVWKNE-DEASIIKQLINLNTIDGITNIGNYKNS 174
P. falciparum  NNNQDILFCNIKDIFGNNKNDKEIHDAILKYGGNERHIIKELRKEKEIGQYNNI--YFN- 177
               *::**** ***: : . : *::* :*: : :* ***:* : : :. .*:..* *:

P. chabaudi    QIYNEIFKKPKENNYVYKSGEGMQNS--YDPISTPEYTSHYIYKG-ADQNSSNYINKTND 231
P. berghei     QNYNETLQKPKENISIYKSGEEIENS--YIPSSTPEYVSHYIYKGRRGENSSNYINKTKD 232
P. falciparum  -DYVNVLNTDPSQNYIYNDMPKITPNNIYNMMNNDQTNHTYLKAPNSLYNNENTIYSSN- 235
               . * : ::. .: :**:. . * ... *: *:.* * .::

P. chabaudi    ILFPNHFNKSTYNDNIKNIYDTPNDTHY--NNSS----TYSNFDHNSMYSTKNNVPFSNP 285
P. berghei     TLFPTYLNNYAYNNNIKNVYDTPNGAHYSSNNNSGSNNAYSNFDHNINNSPKNNVPFSNS 292
P. falciparum  VHYSTYMNNSPTYKNSNNMNHVTNMYASNDLHNSNHFKPHSNAYSTINYDNNNYIYPQNH 295
               :..::*: . .* :*: ...* :.* .:** . . :* : .*

P. chabaudi    NDDKIFQYQQHFQYSRNYIPSPNSEKILYFSCGNKKK 322
P. berghei     DNDKIFH------YSRNYMPSPNSEKMLYFSCGNKKK 323
P. falciparum  TN--IYN-----------RASPGSDQTLYFSPCNQKK 319
               :..*:: .. . : .. **.*:: **** *:**
```

Figure 3. Multiple sequence alignment of the putative N-terminal domain of metacaspases from malaria parasites. The highlighted basic residue at the C-terminus preceding the putative large subunit is well conserved among metacaspases from unicellular organisms. The consensus sequence leading to the identification of PF_0288 is marked with a bar. *P. chabaudi* gi|70947027; *P. berghei* gi|29788140; *P. yoelii* (fragment) gi|82594051; *P. falciparum*, gi|85822146.

Acknowledgements

I thank Hilary Hurd for carefully reading the manuscript and making helpful suggestions. I also gratefully acknowledge Katja Becker for supporting my work on the metacaspases from *P. falciparum* and *S. cerevisiae*.

References

1. Deponte M, Becker K. Plasmodium falciparum—do killers commit suicide? Trends Parasitol 2004; 20:165-69.
2. Hurd H, Carter V. The role of programmed cell death in Plasmodium-mosquito interactions. Int J Parasitol 2004; 34:1459-72.
3. Madeo F, Herker E, Wissing S et al. Apoptosis in yeast. Curr Opin Microbiol 2004; 7:655-60.
4. Madeo F, Herker E, Maldener C et al. A caspase-related protease regulates apoptosis in yeast. Mol Cell 2002; 9:911-17.
5. Uren AG, O'Rourke K, Aravind LA et al. Identification of paracaspases and metacaspases: two ancient families of caspase-like proteins, one of which plays a key role in MALT lymphoma. Mol Cell 2000; 6:961-67.
6. Mottram JC, Helms MJ, Coombs GH et al. Clan CD cysteine peptidases of parasitic protozoa. Trends Parasitol 2003; 19:182-87.
7. Sinden RE, Gilles HM. The malaria parasites. In: Warrell DA, Gilles HM, eds. Essential Malariology, fourth ed. London: Arnold, 2005:8-34.
8. Snow RW, Guerra CA, Noor AM et al. The global distribution of clinical episodes of Plasmodium falciparum malaria. Nature 2005; 434:214-17.
9. Greenwood B, Mutabingwa T. Malaria in 2002. Nature 2002; 415:670-72.
10. Wiesner J, Ortmann R, Jomaa H et al. New antimalarial drugs. Angew Chem Int Ed Engl 2003; 42:5274-93.
11. Attaran A, Roberts DR, Curtis CF et al. Balancing risks on the backs of the poor. Nat Med 2000; 6:729-31.
12. Gardner MJ, Hall N, Fung E et al. Genome sequence of the human malaria parasite Plasmodium falciparum. Nature 2002; 419:498-511.
13. Carlton JM, Angiuoli SV, Suh BB et al. Genome sequence and comparative analysis of the model rodent malaria parasite Plasmodium yoelii yoelii. Nature 2002; 419:512-19.
14. Hall N, Karras M, Raine JD et al. A comprehensive survey of the Plasmodium life cycle by genomic, transcriptomic and proteomic analyses. Science 2005; 307:82-86.
15. Bahl A, Brunk B, Coppel RL et al. PlasmoDB: the Plasmodium genome resource. An integrated database providing tools for accessing, analyzing and mapping expression and sequence data (both finished and unfinished). Nucleic Acids Res 2002; 30:87-90.
16. Crabb BS, Rug M, Gilberger TW et al. Transfection of the human malaria parasite Plasmodium falciparum. Methods Mol Biol 2004; 270:263-76.
17. Thathy V, Menard R. Gene targeting in Plasmodium berghei. Methods Mol Med 2002; 72:317-31.
18. Nicholson DW. Caspase structure, proteolytic substrates and function during apoptotic cell death. Cell Death Differ 1999; 6:1028-42.
19. Vercammen D, van de Cotte B, De Jaeger G et al. Type II metacaspases Atmc4 and Atmc9 of Arabidopsis thaliana cleave substrates after arginine and lysine. J Biol Chem. 2004; 279:45329-36.
20. Watanabe N, Lam E. Two Arabidopsis metacaspases AtMCP1b and AtMCP2b are arginine/lysine-specific cysteine proteases and activate apoptosis-like cell death in yeast. J Biol Chem 2005; 280:14691-99.
21. Bozhkov PV, Suarez MF, Filonova LH et al. Cysteine protease mcII-Pa executes programmed cell death during plant embryogenesis. Proc Natl Acad Sci USA 2005; 102:14463-68.
22. Suarez MF, Filonova LH, Smertenko A et al. Metacaspase-dependent programmed cell death is essential for plant embryogenesis. Curr Biol 2004; 14:R339-40.
23. Al-Olayan EM, Williams GT, Hurd H. Apoptosis in the malaria protozoan, Plasmodium berghei: a possible mechanism for limiting intensity of infection in the mosquito. Int J Parasitol 2002; 32:1133-43.
24. Szallies A, Kubata BK, Duszenko M. A metacaspase of Trypanosoma brucei causes loss of respiration competence and clonal death in the yeast Saccharomyces cerevisiae. FEBS Lett 2002; 517:144-50.
25. Singh GP, Chandra BR, Bhattacharya A et al. Hyper-expansion of asparagines correlates with an abundance of proteins with prion-like domains in Plasmodium falciparum. Mol Biochem Parasitol 2004; 137:307-19.
26. Riedl SJ, Renatus M, Schwarzenbacher R et al. Structural basis for the inhibition of caspase-3 by XIAP. Cell 2001; 104:791-800.
27. Guex N, Peitsch MC. SWISS-MODEL and the Swiss-PdbViewer: an environment for com-parative protein modeling. Electrophoresis 1997; 18:2714-23.

CHAPTER 9

Cell Death in Trichomonads

Marlene Benchimol*

Abstract

Programmed cell death (PCD) is not confined to mammals and is extremely widespread, possibly universal, in multicellular animals. It is now evident that PCD also occurs in single-celled organisms and it is an important feature of host-pathogen relationships. Protists are capable of eliciting an apoptotic response in several circumstances. Studies are in course in order to establish whether these organisms share some or all of the effectors and regulators common to multicellular PCD or have evolved their own divergent pathways. Trichomonads are amitochondrial protists that inhabit different ecological niches. Among them *Tritrichomonas foetus*, a cattle parasite and *Trichomonas vaginalis*, a human parasite, are the most important because they cause trichomoniasis, a sexually transmitted disease. They do not possess mitochondria, but harbor another type of membrane-bounded organelle, an unusual anaerobic energy-producing organelle called hydrogenosome. Studies of cell death in trichomonads are under way in order to establish whether the hydrogenosome could represent an alternative to mitochondria whether these organisms possess all caspase activities and which conditions lead trichomonads to cell death. In these organisms the known "mitochondrial cell death machinery" is supposed to be distinct from mitochondrial eukaryotes. The presence of a cell death program in trichomonads suggests the existence either of a dependent or independent caspase-like execution pathway in such organisms. Dramatic changes in trichomonad morphology are observed when the cells are under stress, such as after drug treatment and nutrient depletion. These changes include intense plasma membrane and nuclear envelope blebbing, nucleus fragmentation, an abnormal number of oversized vacuoles and altered hydrogenosomes. DNA fragmentation, exposure of phosphatidylserine (PS) in the outer leaflet of the plasma membrane, hydrogenosomal membrane potential dissipation, are features observed in apoptotic cells. Trichomonads also present autophagic processes, observed when altered hydrogenosomes, misshapen flagella, abnormal cellular elements and tubulin precipitates are located in autophagic vacuoles, which are limited by a double or multiple concentric membrane. In all stress situations, trichomonads form pseudocysts, cells with internalized flagella. Different forms of cell death, such as apoptosis, autophagy and necrosis have been shown to exist in trichomonads and so the possibility of the existence of different pathways to cell death in trichomonads is raised.

Introduction

Trichomonads are amitochondrial protists that inhabit different ecological niches. Among them, two species are very important because they are parasites of the urogenital tract and cause trichomonosis, a sexually transmitted disease: (1) *Tritrichomonas foetus*, a cattle parasite (Figs. 1, 2), and (2) *Trichomonas vaginalis*, a human parasite. They do not possess mitochondria, but harbor an unusual anaerobic energy-producing organelle called hydrogenosome (Figs. 1, 11) due its capacity to produce molecular hydrogen. Since mitochondria participates in cell death processes in higher

*Marlene Benchimol—Rua Jornalista Orlando Dantas 59, Botafogo, Rio de Janeiro, RJ. Brazil. CEP 222-31-010. Email: marleneben@uol.com.br

Programmed Cell Death in Protozoa, edited by José Manuel Pérez Martín.

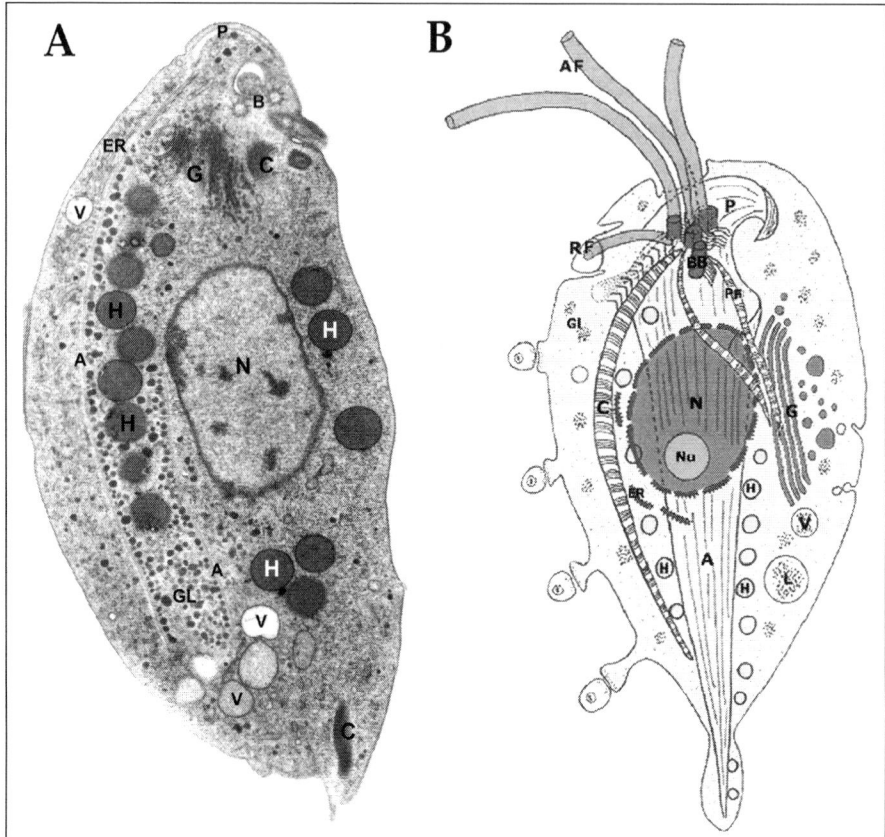

Figure 1. Thin section (A) and schematic diagram (B) of *Tritrichomonas foetus* showing the main cell structures. A) General view of *T. foetus* in a longitudinal routine thin section. An oval nucleus (N), small-sized vacuoles (V), hydrogenosomes (H), endoplasmic reticulum (ER), pelta (P), basal bodies (B), Golgi complex (G), axostyle (A), costa (C) and glycogen granules (GL) are seen. Bar 1 μm. B) AF, anterior flagella; Ax, axostyle; BB, basal body; C, costa; ER, endoplasmic reticulum; G, Golgi; H, hydrogenosomes; L, lysosome; N, nucleus; Nu, nucleolus; P, pelta; PF, parabasal filament; RF, recurrent flagellum; V, vacuole. (After Benchimol, 2004.)

eukaryotes, cells without mitochondria, but possessing hydrogenosomes led researchers to debate whether this organelle could be a natural substitute of mitochondria during apoptosis.

Programmed cell death (PCD) is not confined to mammals and is extremely widespread, possibly universal, in multicellular animals. It is now evident that PCD also occurs in single-celled organisms and it is an important feature of host-pathogen relationships. Protists are capable of eliciting an apoptotic response in several circumstances. Studies are in course in order to establish whether these organisms share some or all of the effectors and regulators common to multicellular PCD or have evolved their own divergent pathways.

The existence of PCD in a variety of single-celled eukaryotes of distinct phylogenic origins indicated a functional role of this process in the biology of such organisms. Among trichomonads the main publications in cell death are those in *Trichomonas vaginalis*[1] and *Tritrichomonas foetus*.[2-5] Several inductors of cell death in trichomonads have been used and the main results are presented (see Tables 1-3).

Table 1. Morphological features presented by Trichomonads during apoptotic cell death

Apoptotic Bodies (zooids)	Endoplasmic Reticulum (dilated cisternae)	Plasma Membrane Blebbing	Nuclear Envelope Blebbing	Chromatin Condensation	Cytoplasmic Vacuolation	Nucleus Fragmentation	Abnormal Hydrogenosomes	Hydrogenosomes with Nucleoid	Flagella Internalization (Pseudocyst)
+	+	+	+	+	+	+	+	+	+

Table 2. Inductors to cell ceath in Trichomonads

Microtubule Interfering Drugs	Microfilament Interfering Drugs	Metabolism Interfering Drugs	Starvation	Others
Colchicine[15]	Cytochalasin B[3]	Metronidazole[1], Benchimol, unpublished	Serum deprivation[2,7]	Fibronectin[3]
Nocodazole[7,15]	Cytochalasin D[7]	Staurosporine[1]	—	Lectins (HPA, WGA) (Benchimol, unpublished)
Griseofulvin[5]	—	Etoposide[1] (ETO)	—	—
Taxol[7,15]	—	Doxorubicin[1] (DOXO)	—	—
—	—	H_2O_2[4]	—	—
—	—	Hydroxyurea[2];	—	—
—	—	zinc sulfate[2,48]	—	—

Table 3. Characteristics of cell death in Trichomonads

Motility	PS Exposure	TUNEL	Autophagy	Necrosis	Pseudocyst Formation	Caspase	Hydrogenosomal Membrane Potential Dissipation
Decreased[1,4-5,7]	Phosphatidyl serine in the outer plasma membrane leaflet[1,5]	DNA fragmentation[1]	Hydrogenosome[2], cytosol portions and microtubular structures[5]	Evidence of secondary necrosis and cytolysis (Benchimol, unpublished observations)	Flagella internalization[7-8]	Presence of a caspase-like pathway[1,4], and a caspase-independent pathway[5]	DiOC$_6$ marker[1]

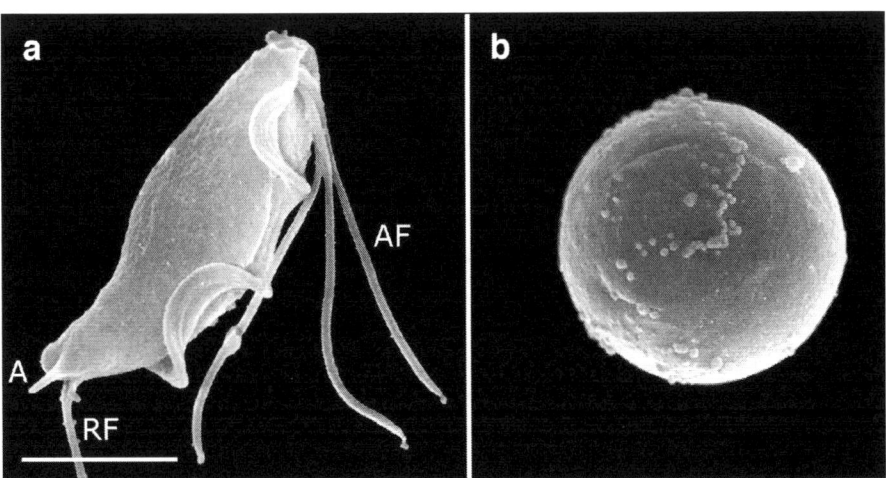

Figure 2. Scanning electron microscopy of the two forms of trichomonads: trophozoite (a) and pseudocyst (b). Note that the trophozoite presents externalized flagella and has a pear-shape, whereas the pseudocyst presents internalized flagella and the cell becomes rounded. Pseudocyst exists when the cells are under stress. A, axostyle; AF, anterior flagella; RF, recurrent flagellum. Bar = 1 μm. (After Benchimol, 2004.)

Trichomonas Structure

T. foetus has three anterior flagella whereas *T. vaginalis* has four and both have a recurrent flagellum that runs toward the posterior region of the cell adhering to the cell body, forming an undulating membrane. The mastigont system also comprises several skeletal structures including the costa (Fig. 1) that underlies the recurrent flagellum along the undulating membrane and the parabasal filament that supports a single Golgi complex.[6] A nondividing *T. foetus*, grown in axenic medium, is characterized by a pear-shaped body, measuring approximately 16 μm long and 7 μm wide, with one anterior nucleus, a pelta supporting the flagellar canal, a single ribbon of microtubules forming the axostyle, which runs from the basal body to the cell tip and hydrogenosomes lining the axostyle (Figs. 1, 11).

Pseudocyst

Pseudocyst formation refers to the morphological transformation of trichomonads into compact, less-motile forms, without a true cyst wall and exhibiting internalized flagella (Fig. 2B). It represents a response to environmental stress. Trophozoite forms of these organisms, which are polar and flagellated, become rounded and devoid of external flagella upon pseudocyst formation. The flagella can be seen to reside within the cell body of the pseudocysts.[7,8] This form may lead to a better understanding of how pathogenic trichomonads survive and how they are effected by therapeutic drugs. When trichomonads are under drug treatment or nutrient depletion, cells transform in pseudocysts, which can be either a temporary/reversal form or a phase just before trichomonad cell death.

Motility

When trichomonads are stressed by drugs or nutrient depletion, the cells exhibit a significant inhibition in motility (Fig. 12). The first sign of stress is flagella internalization (Figs. 2B, 6). This process can be reverted in some situations, indicating that the cells had not fully reached the death pathway.[7] These authors described an in vitro system of reversible pseudocyst formation in *T. foetus* treated with a variety of agents.

Detection of Apoptosis-Like Death in Trichomonads

One of the most common methods for the detection of apoptosis-like death in protozoa is nuclear DNA fragmentation (Fig. 5), which can be revealed using two different approaches: one is agarose gel electrophoresis where ladder formation of bands are observed and the other is the TUNEL label technique where DNA fragmentation can be revealed by fluorescence microscopy or flow cytometry (Fig. 15), this latter case is more usual to quantify the fragmentation. Transmission electron microscopy of thin sections has also been used to show condensation of the nuclear chromatin (Fig. 13) and blebs of the plasma membrane as shown in Figures 6 and 10. This technique also permits visualization of cytoplasmic vacuolation, as well as membrane breakdown, features observed in autophagic cell death and necrosis, respectively (Figs. 11c, 14). Exposition of phosphatidylserine on cell surfaces (Fig. 15), evaluated by flow cytometry or fluorescence microscopy using labeled annexin-V, has also been used as a criterion to identify apoptotic in trichomonads cells.[1,4,5]

Previous workers[1,4] have shown the presence of an execution program in the amitochondriate *T. vaginalis* and *T. foetus*, respectively. In both works it was suggested that there is the existence of a caspase-like pathway for cell death in trichomonads. It is well established that reactive oxygen species (ROS) generated by antiparasitic agents or macrophages can induce cell death in intracellular parasites[9] and are therefore important regulators of protozoan infection. The cytotoxic effects and the induction of cell death by hydrogen peroxide (H_2O_2) on *T. foetus* have been demonstrated previously.[4] The authors tested whether in this trichomonad the cell death program involves the participation of caspase-3 (Figs. 3, 4), one of the key executioners of apoptosis in several other organisms (for a review see ref. 10). It has been shown in different cell types that chromatin condensation involves the participation of acinus, a caspase-3-activated factor.[11] Moreover, the nuclear fragmentation itself occurs due to the cleavage of lamins by caspase-6.[12,13] However, the presence of these last molecules in trichomonads is unknown. One group[1] has reported the possible presence of caspase-3 using an anti-caspase-3 antibody, but unfortunately this result has not been documented by the authors. However, other group[4] obtained indication that at least a caspase-3-like protein, if not caspase-3 itself, is present and activated during H_2O_2-induced PCD in this trichomonad (Figs. 3, 4).

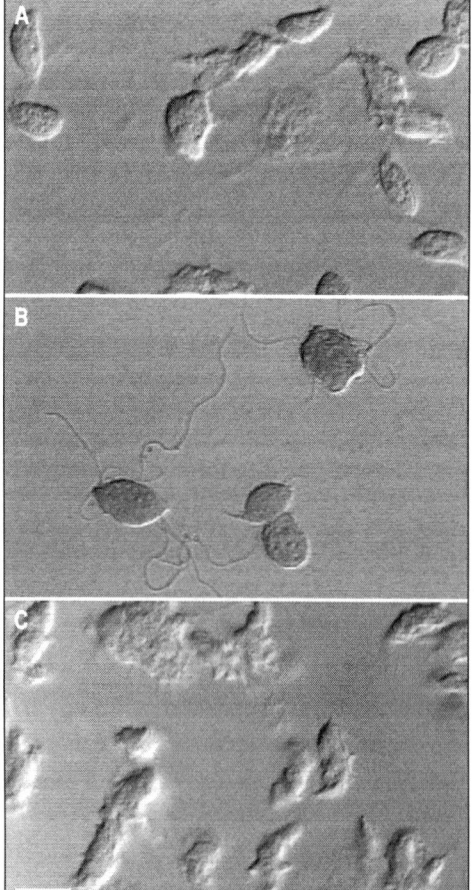

Figure 3. DIC microscopy of *T. foetus* after immunocytochemistry for activated caspase-3. A) Untreated cells. B) Cells treated with 0.5 mM H_2O_2 for 4 h (positive control). C) Cells treated with 50 µg/ml of griseofulvin for 24 h. Note that griseofulvin treated cells do not present any labeling, in a similar way to the untreated cells, whereas cells treated with H_2O_2 (Fig. 3B) exhibit an intense positive labeling. Bar 10 µm. (After Mariante et al. 2003.)

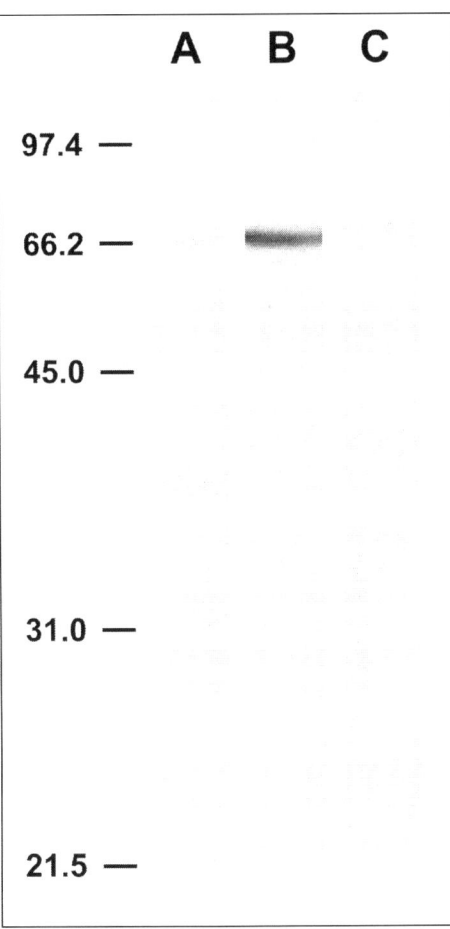

Figure 4. Western blotting analysis of *T. foetus* whole-cell protein extracts with the CM1 antibody. Lane A, untreated cells. Lane B, cells treated with 0.5 mM H_2O_2 for 4 h. Lane C, negative control provided by the omission of the CM1 antibody. The protein molecular weight is expressed in kDa. (After Mariante et al 2003.)

Second Pathway

Recently, our group[5] raised the possibility that trichomonads can present a different execution pathway, apparently independent of caspase-3. Unlike hydrogen peroxide,[4] griseofulvin led cells to death without the involvement of caspase-3.

Transmembrane Potential Disruption

The production of ROS in mitochondrial parasites treated with H_2O_2 is accompanied by the loss of the mitochondrial membrane potential ($\Delta\Psi$) and an increase in the Ca^{2+} pool in the cytoplasm.[14] All these changes are usually associated with cell death by apoptosis. In order to verify if the amitochondriate parasite *T. vaginalis* exhibits loss of the hydrogenosomal membrane potential ($\Delta\Psi$), some authors[1] measured the membrane potential disruption using a $DiOC_6$ marker in a flow cytometry experiment. They observed a weak but significative membrane potential dissipation when *T. vaginalis* was treated with pro-apoptotic drugs.

Plasma Membrane Blebbing and Apoptotic Bodies

Some features commonly found in cell death of higher eukaryotic cells are also observed in trichomonads under stress. These include an intense plasma membrane blebbing[4,5,15] (Fig. 6), shedding of apoptotic bodies[15] (Fig. 10) and also an interesting nuclear envelope blebbing[5] (Fig. 8). Apoptotic bodies or zooids were easily seen when lectins such as HPA or WGA were used in living trichomonads (Fig. 10) (Benchimol, unpublished).

Hydrogenosomes

The hydrogenosome is a key organelle in trichomonads (Figs. 1, 11). It contains enzymes that participate in the metabolism of pyruvate and is the site of ATP and molecular hydrogen formation (for a review see ref. 16). The origin of the hydrogenosome has been a subject of intense discussion, since, like mitochondria, it is enveloped by two membranes,[17] divides autonomously by fission,[18] imports proteins post-translationally,[19] possesses cardiolipin,[20] and produces ATP.[21] However, differently from mitochondria, the hydrogenosome of trichomonads seems to lack a genome, a respiratory chain, cytochromes, the F0-F1 ATPase, the tricarboxylic acid cycle and oxidative phosphorylation.[16,22] Several hypotheses have been raised to explain the origin of the trichomonads hydrogenosome: an independent endosymbiosis of an anaerobic eubacterium with a eukaryotic

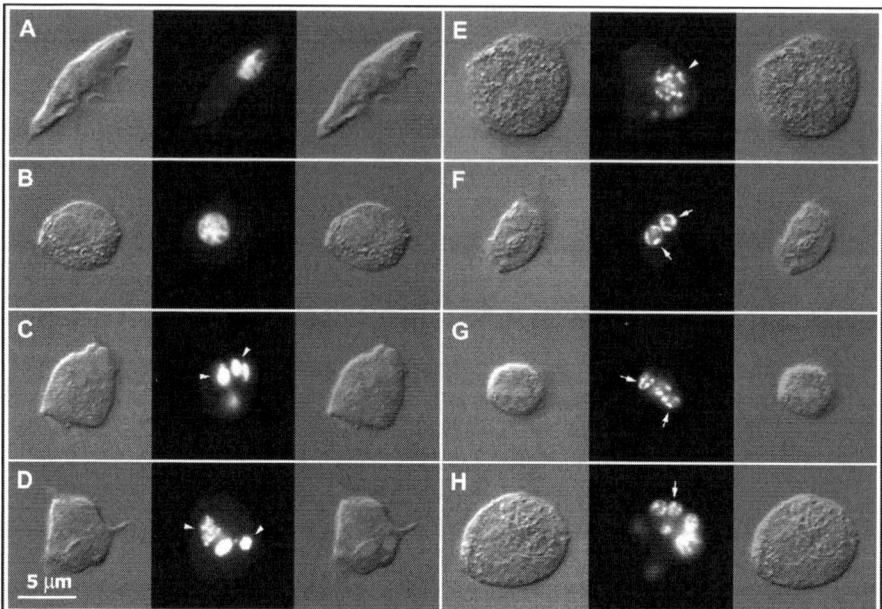

Figure 5. Fluorescence microscopy of *T. foetus* stained with 5 mg/ml DAPI. A) Control cell with no H_2O_2 treatment. B-H *T. foetus* cells treated with 4 mM H_2O_2 for 10 min (B,C), 20 min (D,H) and 1 h (E-G). Nuclear fragmentation (C-E, arrowheads) and peripheral masses of heterochromatin (F-H) are seen. (After Mariante et al 2003.)

host or the conversion of an established mitochondrion adapted to an anaerobic lifestyle. The common hypothesis for the origin of mitochondria and hydrogenosomes was refined to state that both organelles evolved from a common progenitor structure present in eukaryotes before the advent of true mitochondria or hydrogenosomes.[23,24]

It is well established that mitochondria play a pivotal role during cell death of many different cell types, including unicellular organisms.[25-28] The release of proteins from mitochondria into the cytosol is one of the most common mechanisms in promoting and regulating cell death in eukaryotic cells.[29,30] At least through the release of cytochrome c, mitochondria seem to play a pivotal role in the apoptotic pathways.[31,32] Since trichomonads do not seem to possess such a ubiquitous mechanism, the pathways that lead to cell demise in these unicellular parasites remain to be elucidated. Figure 16 shows a proposed model for possible execution of cell death in trichomonads. There are at least two primary apoptotic mechanisms. The mitochondrion-dependent pathway activated by the release of cytochrome c from the mitochondrial intermembrane space and the activation of surface "death receptors" resulting in a signaling cascade.[33] At present, most of the studies carried out in protozoa point to the first mechanism although the second one can not be eliminated, especially in protozoa, which do not have mitochondria.

Hydrogenosomes present evident alterations when trichomonads are submitted to stress (Fig. 11). In trichomonads treated with drugs the hydrogenosomes are clearly affected: they become abnormal, presenting various different bizarre shapes and a higher matrix electron density (see Table 2 for references). When the hydrogenosome is dead it presents a nucleoid, a dense precipitate in its matrix (Table 1).

Further studies are necessary to improve our knowledge about the role of hydrogenosomes during trichomonads PCD. It is possible that hydrogenosomes may be involved in the cell death pathway of trichomonads in a similar mode to mitochondria, by releasing hazardous proteins into the

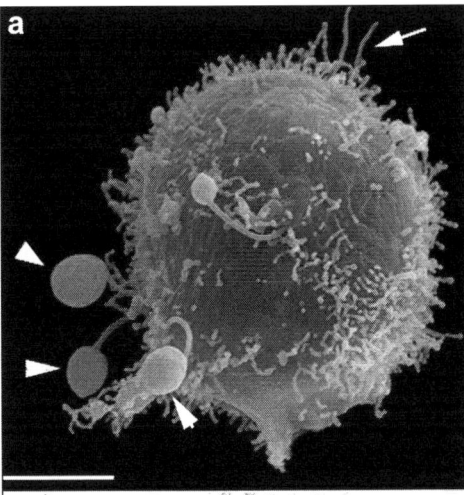

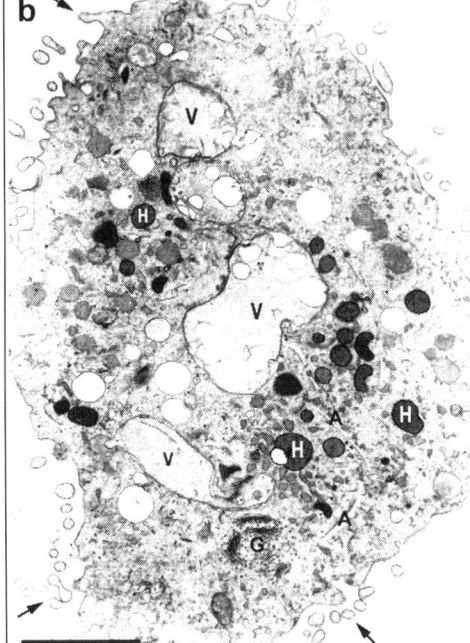

Figure 6. Scanning (a) and transmission (b) electron microscopy of *T. foetus* observed after 50 μg/ml griseofulvin treatment for 24 h. a) An intense membrane blebbing is observed and all the flagella (F) are internalized. Note that the flagella ends in large blebs (arrowheads). Bar = 2 μm. b) Thin section of a cell presenting an intense membrane blebbing (*arrows*) and cytoplasmic vacuolization (V). Hydrogenosomes (H) exhibit high electron-dense contents. The Golgi complex (G) is maintained. The axostyle-pelta complex (A) became fragmented after drug treatment. Bar 3 μm. (After Mariante et al 2005.)

cytosol upon damaging stimuli. It is known that the defensive mechanisms of *T. foetus* against the toxic products of O_2 reduction involves the participation of at least three enzymes: superoxide dismutase,[34] peroxidases (e.g., ascorbate peroxidase[35]) and catalase.[35-37] The activity of the first two enzymes was found not only in the cytosol, but also inside the hydrogenosomes. Since these enzymes participate in detoxifying O_2 products, it would be reasonable to imagine that the presence of H_2O_2 could lead to the activation of superoxide dismutase and/or ascorbate peroxidase, resulting in degradation of H_2O_2 and precipitation of its products in the hydrogenosome matrix.[4]

Endoplasmic Reticulum and Golgi Complex Behavior

The endoplasmic reticulum is also affected when the cell presents signs of cell death.[38,2,3] The main changes observed include reticulum proliferation, lumen dilatation and organelle enlargement (Fig. 13). It has been shown that Golgi fragments during the mitotic and apoptotic events in higher eukaryotic cells.[39] Differently, the Golgi complex of trichomonads is not morphologically affected when these cells are under drug treatment (Fig. 6b). This is a very interesting observation since the Golgi of trichomonads has an uncommon behavior during mitosis.[6] It has been shown that, in mammalian cells, many proteins associated with Golgi, such as golgin-160, responsible for maintaining the Golgi structure,[40] GRASP65, a protein involved in the stacking of Golgi cisternae[41] and p115, a vesicle-tethering protein,[42] are cleaved by caspases during apoptotic cell death. The cleavage of such proteins induces the fragmentation of the organelle. Thus, it would be reasonable to suggest that in trichomonads the involvement of similar mammalian Golgi proteins does not take place, neither in the mitotic nor during apoptotic events.

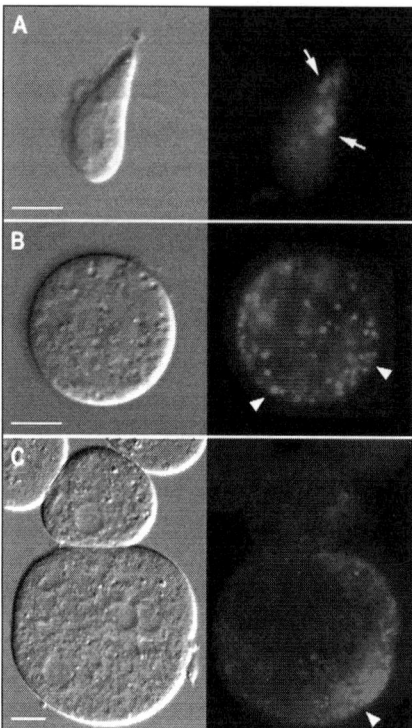

Figure 7. DIC and fluorescence microscopy of *T. foetus* stained with 20 μg/ml acridine orange. A) Control cell presenting intense labeling in lysosomes which are posteriorly located (*arrows*). B,C) Cells treated with 50 μg/ml griseofulvin for 24 h. Giant cells presenting several stained vacuoles randomly distributed (arrowheads) are seen. Bars 5 μm. (After Mariante et al 2005.)

Lysosomes and Vacuoles

Lysosomes and vacuoles are normally seen in the posterior region of the cell in nonstressed conditions in trichomonads. However, during cell-stress the lysosomes in trichomonads become much larger and are distributed throughout the cell cytoplasm. Intense vacuolization and the formation of myelin-like figures are frequently observed during cell death in trichomonads (Figs. 6b, 7, 9).

Cytoskeleton

When trichomonads are under stress cytoskeletal structures such as the axostyle are affected. As the cell becomes rounded, the axostyle is seen fragmented. The axostyle is a microtubular structure that does depolymerize, neither during mitosis[43] nor during drug treatment.[15] The axostyle is broken up in fragments, allowing the cell to be rounded-up. Drugs affecting the cytoskeleton such as cytochalasins (Benchimol, unpublished observations), taxol, colchicine or nocodazole have shown similar cell behavior.[15]

Nucleus Behavior

The apoptotic nucleus in trichomonads presented some differences and similarities when compared with mammalian apoptotic cells. Peripheral masses of heterochromatin, nuclear fragmentation and an uncommon condensation patterns are observed in several, but not in all, apoptotic trichomonads[1,4,5] (Figs. 5, 8, 13a). Interestingly, the nuclear envelope also exhibited membrane blebbing (Fig. 8). DNA fragmentation and TUNEL positive cells were observed when trichomonads were drug-treated.[1,4,5]

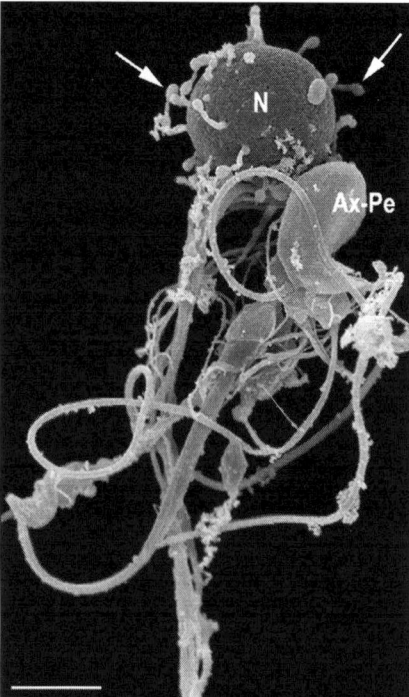

Figure 8. Field emission scanning electron microscopy of *T. foetus* under griseofulvin treatment followed by permeabilization, exposing the nucleus and axostyle-pelta complex (Ax-Pe). The nucleus presents blebbing on the nuclear envelope (arrows). Bar 20 μm. (After Mariante et al 2005.)

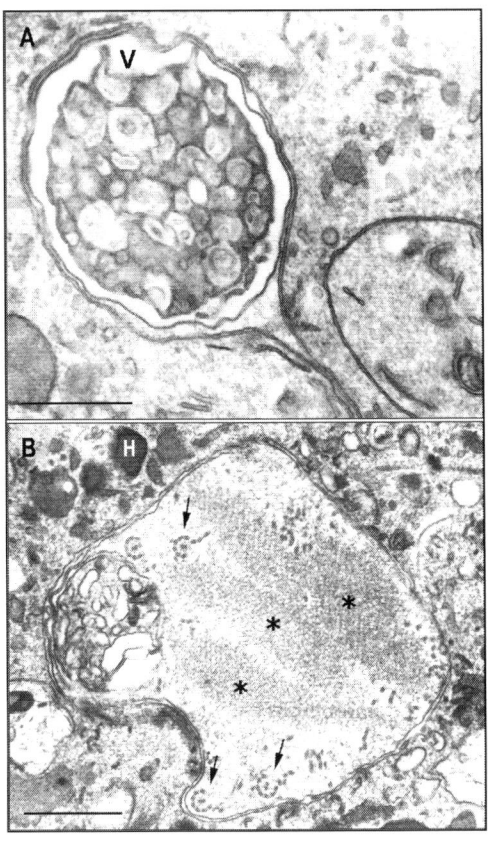

Figure 9. *T. foetus* vacuoles (V) after griseofulvin treatment for 24 h. Membranous profiles are found within most of the vacuoles (A,B). Several compartments were seen fulfilled with abnormal flagella axonemes (B, arrowheads) and microtubule-like deposits (B *asterisks*). H, hydrogenosome. Bars 700 nm (A), 800 nm (B). (After Mariante et al 2005.)

Induction of Cell Death in Trichomonads

Several cell death inductors have been used in trichomonads in order to better understand CD in these parasite protists (Table 2). Among them, staurosporine, etoposide and DOXO (doxorubicin) were used,[1] whereas hydrogen peroxide[4] (Figs. 3-5), hydroxyurea[44] (Fig. 11) and drugs affecting the cytoskeleton, such as nocodazole, colchicine, cytochalasins and griseofulvin (Figs. 6-9, 12, 15) have been used by our group[2-7,15] (see Table 2). In addition, starvation, metronidazole and fibronectin have also been tested.[3] Lectins were also shown to provoke apoptosis in *T. foetus* (Benchimol, unpublished); cells exhibited intense blebbing and apoptotic bodies shedding (Fig. 10).

In order to have a better understanding of the behavior of trichomonad organelles when in apoptosis, other drugs have been used, such as acridine orange (Fig. 7). In addition, to verify the presence of autophagic vacuoles, didansylcadaverine (DDC) and 3-Methyladenine (3-MA) were used. The incorporation of DDC was via autophagy and 3-MA was used as an autophagic inhibitor.

Autophagy

Programmed cell death occurs in various forms, such as apoptosis, autophagy and others (Fig. 11c).

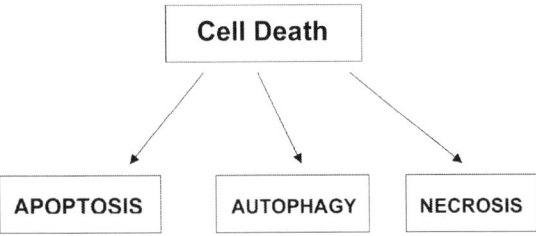

Autophagy is a type of non-apoptotic death that has been classified as a programmed necrosis or autophagic cell death.[45-47] In programmed necrosis there is a cellular signaling pathway leading

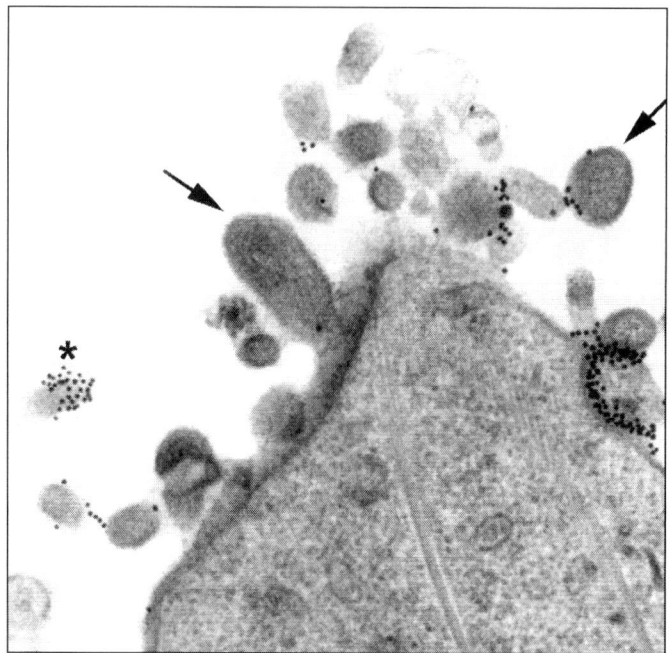

Figure 10. Electron microscopy of *T. foetus* incubated with HPA (*Helix pomatia*), a gold conjugated lectin. Signs of cell death such as membrane blebbing (arrows) is seen. Asterisk points to gold particles in an apoptotic body. Bar 200 nm.

to death in response to specific cues and not due to an "accident". The name autophagy means, literally, to eat oneself. In some cases trichomonads can die either by programmed necrosis or by autophagy, in a process independent of caspase activation or other features of apoptosis. Both types of cell death may represent the end of a wide spectrum of physiological cell death. However, there are circumstances where apoptosis and autophagy may pertain to the same pathway of programmed death. The process of autophagy has been studied in *Trichomonas* under nutrient deprivation, drug treatment such as hydroxyurea[2] and zinc sulfate[48] and also under normal conditions.[2] Apparently normal hydrogenosomes, as well as giant, abnormal and those hydrogenosomes presenting internal cristae are the are the most affected structures observed during the autophagic process.[2] During autophagy, hydrogenosomes or other cell structures are first sequestered away from the remaining cytoplasm and then degraded within lysosomes. The autophagic vacuoles were limited by a double or multiple concentric membranes (Fig. 11). The first event observed was the rough endoplasmic reticulum surrounding and enclosing the hydrogenosome forming an isolation membrane. Next, the autophagic vacuole fuses with lysosomes and also with other vacuoles containing the organelles. Large vacuoles containing many hydrogenosomes or other altered cell components (Figs. 9, 11) are formed and then partially degraded in lysosomes. The process of autophagy has been observed when trichomonads are under drug treatment or even in normal cells[2] and constitutes a new aspect of cell death in trichomonads that deserves further in-depth studies.

Necrosis

In contrast to apoptosis, necrosis appears to be a passive form of cell death with more similarities to a fatal accident than suicide. In many cases necrosis is the result of a bioenergetic catastrophe, after a complete ATP depletion, probably initiated mainly by cellular "accidents" such as toxic

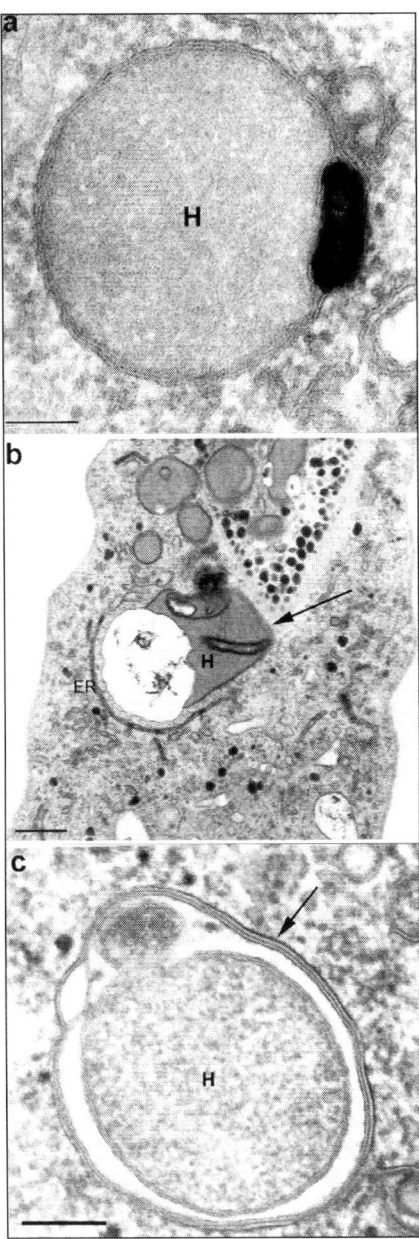

Figure 11. Hydrogenosomes (H) under normal conditions (a) and after drug treatment (b-c). a) Thin section of a routine hydrogenosome where the double membrane and a peripheral vesicle containing a calcium deposit is seen. Bar 100 nm. b,c) *T. foetus* hydrogenosome after drug treatment with hydroxyurea. b) An abnormal hydrogenosome (H) is surrounded by profiles of endoplasmic reticulum (ER). Bar 500 nm. c) A hydrogenosome (H) during an autophagic process, surrounded by a double-bounded membrane vacuole (arrow) is seen. Bar 100 nm. (After Benchimol, 1999.)

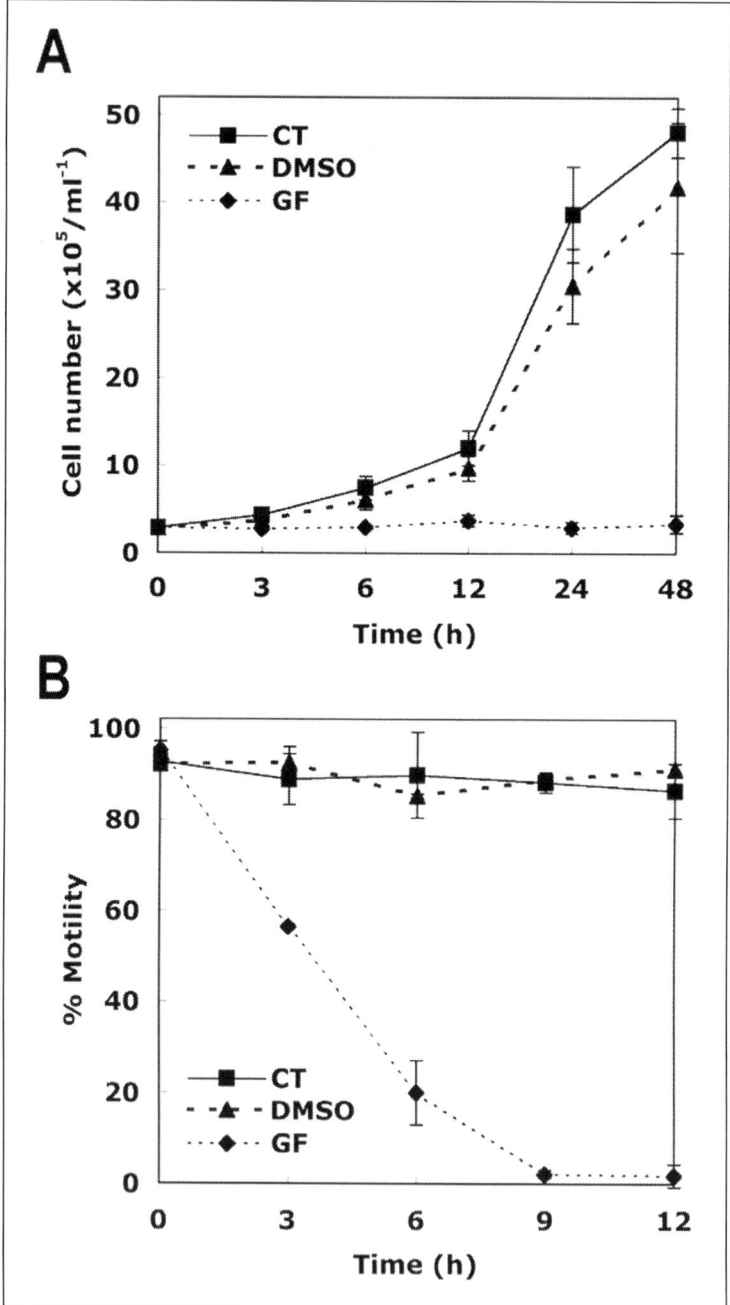

Figure 12. Griseofulvin effects on *T. foetus* growth (A) and motility (B). The drug was added at 50 μg/ml with different time intervals. Legend: CT control with no drug treatment; DMSO control with 0.5% dimethyl sulfoxide, the fungicide diluting agent; GF 50 μg/ml griseoful-vin. Note the inhibitory effects of the compound within the first hours of treatment. (After Mariante et al 2005.)

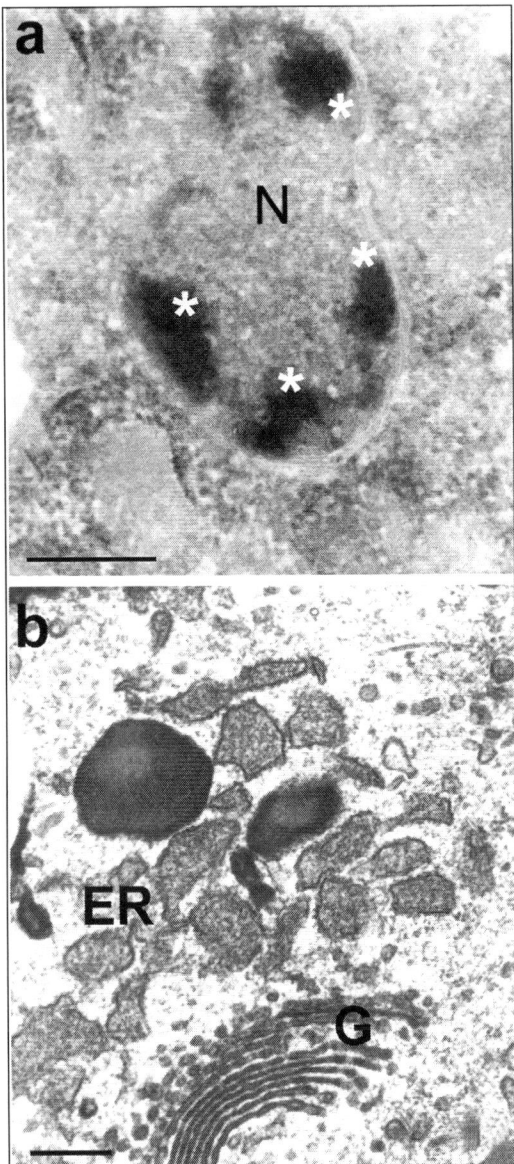

Figure 13. *T. foetus* presenting cell death characteristics after nutrient depletion: nuclear fragmentation with peripheral masses of heterochromatin (a, asterisks) and proliferation, lumen dilatation and enlargement of the endoplasmic reticulum (ER) (b). G, Golgi. Bars 500 nm (a); 200 nm (b).

insults or physical damage. In necrosis, differently from apoptosis, a breakdown of the plasma membrane and release of the cellular contents are observed (Fig. 14) (Benchimol, unpublished). It should also be mentioned that some authors consider the non-apoptotic programmed cell death as a new process designated as paraptosis.[49] Since intense vacuolization was observed during *T. vaginalis* cell death this nomenclature was used by Chose.[1]

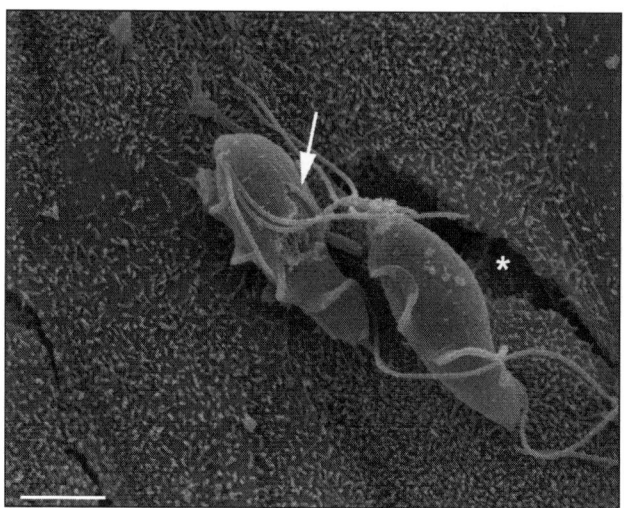

Figure 14. Scanning electron microscopy of interaction of *T. foetus* (T) with bovine culture of oviduct. Asterisk points to an injury in the epithelial culture. Note that two *T. foetus* are tightly adhered to an epithelial cell, but while one is normal, the other trichomonad (arrow) is damaged to death, presenting an evident breakdown of its plasma membrane (necrosis). Bar 3 μm.

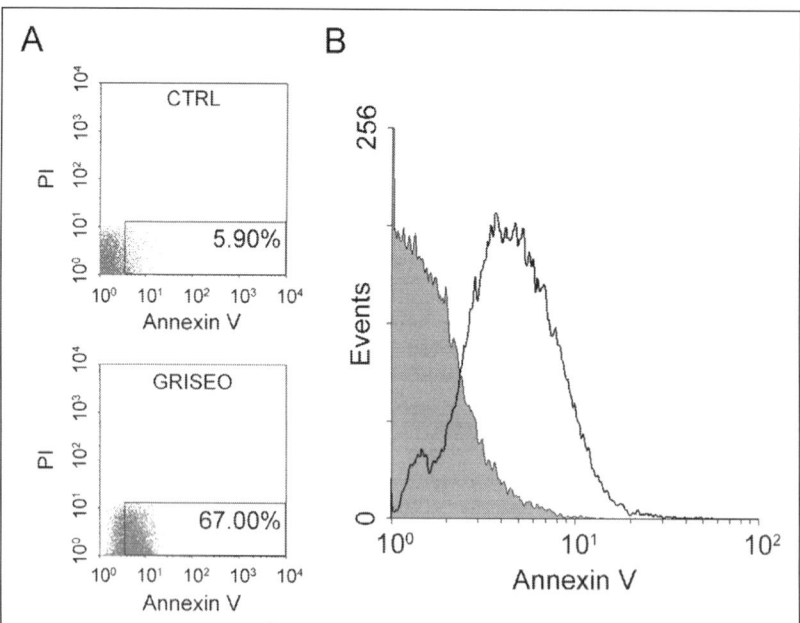

Figure 15. Flow cytometry analysis of trichomonad cell death. A) *T. foetus* was incubated in the absence (CTRL) or presence (GRISEO) of griseofulvin (50 μg/ml) for 24 h and PS exposure was analyzed. Percentages of positive cells (squares) are indicated. B) Fluorescence intensity of griseofulvin treated parasites (open histogram). Filled histogram represents basal cell fluorescence in the absence of the drug. (After Mariante et al 2005.)

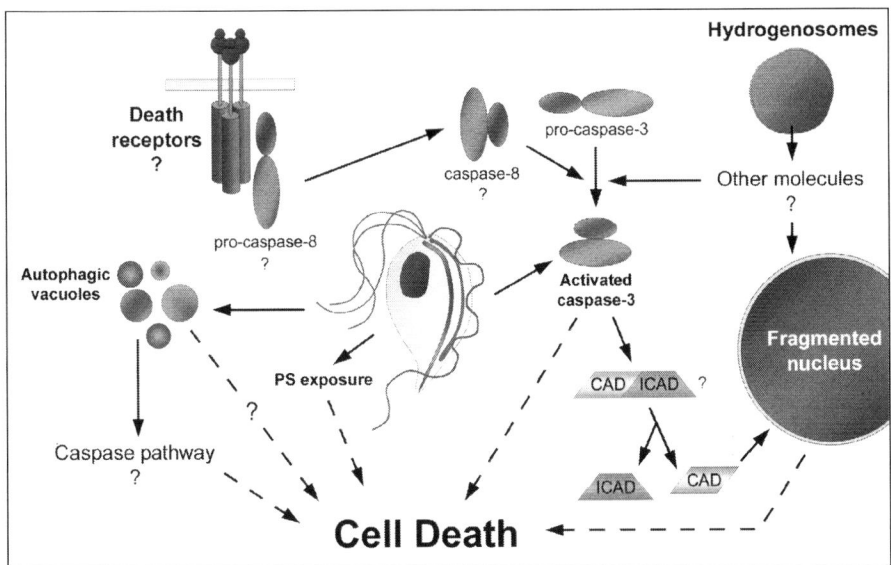

Figure 16. Proposed model for possible execution pathways during *T. foetus* cell death, which includes the presence of possible death receptors, hydrogenosome participation, autophagy and nucleus fragmentation. Abbreviations: PS phosphatidylserine, CAD caspase-activated DNase, ICAD inhibitor of CAD. (After Mariante et al 2005.)

During our experiments of coculture of *T. foetus* with cell cultures we have observed damage and death in trichomonads. Many cells were found lysed with evident signs of injury and necrosis (Fig. 14). This injury to trichomonads would explain the infection clearance observed in contaminated cows.

Factors liberated by the host-cells are under study to clarify how trichomonads die during the process of infection.

Acknowlegements

This work was supported by CNPq (Conselho Nacional de Desenvolvimento Científico e Tecnológico), PRONEX (Programa de Núcleo de Excelência), FAPERJ (Fundação Carlos Chagas Filho de Amparo à Pesquisa do Estado do Rio de Janeiro), FENORTE (Fundação do Norte Fluminense) and AUSU (Associação Universitária Santa Úrsula).

References

1. Chose O, Noel C, Gerbod D et al. A form of cell death with some features resembling apoptosis in the amitochondrial unicellular organism Trichomonas vaginalis. Exp Cell Res 2002; 276:32-39.
2. Benchimol M. Hydrogenosome autophagy: an ultrastructural and cytochemical study. Biol Cell 1999; 91:165-174.
3. Benchimol M. Hydrogenosome morphological variation induced by fibronectin and other drugs in Trichomonas vaginalis and Tritrichomonas foetus. Parasitol Res 2001; 87:215-222.
4. Mariante RM, Guimarães CA, Linden R et al. Hydrogen peroxide induces caspase activation and programmed cell death in the amitochondrial Tritrichomonas foetus. Histochem Cell Biol 2003; 120:129-141.
5. Mariante RM, Vancini R, Benchimol M. Cell death in Trichomonads: new insights. Histochem Cell Biol 2005; 5:1-12.
6. Benchimol M, Mariante RM, Ribeiro KC et al. Structure and Division of the Golgi complex in Trichomonas vaginalis and Tritrichomonas foetus. Eur J Cell Biol 2001; 80:593-607.
7. Granger BL, Warwood SJ, Benchimol M et al. Transient invagination of flagella by Tritrichomonas foetus. Parasitol Res 2000; 86:699-709.

8. Pereira AN, Ribeiro KC, Benchimol M. Pseudocysts in Trichomonads- new insights. Protist 2003; 154:313-327.

9. Mauel J, Schnyder J, Baggiolini M. Intracellular parasite killing induced by electron carriers. II. Correlation between parasite killing and the induction of oxidative events in macrophages. Mol Biochem Parasitol 1984;13:97-110.

10. Cohen GM. Caspases: the executioners of apoptosis. Biochem J 1997; 326:1-16.

11. Sahara S, Aoto M, Eguchi Y et al. Acinus is a caspase-3-activated protein required for apoptotic chromatin condensation. Nature 1999; 401:168-173.

12. Lazebnik YA, Takahashi A, Moir RD et al. Studies of the lamin proteinase reveal multiple parallel biochemical pathways during apoptotic execution. Proc Natl Acad Sci USA 1995; 92:9042-9046.

13. Takahashi A, Alnemri ES, Lazebnik YA et al. Cleavage of lamin A by Mch2 alpha but not CPP32: multiple interleukin 1 beta-converting enzymerelated proteases with distinct substrate recognition properties are active in apoptosis. Proc Natl Acad Sci USA 1996; 93:8395-8400.

14. Mukherjee SB, Das M, Sudhandiran G et al. Increase in cytosolic Ca^{2+} levels through the activation of nonselective cation channels induced by oxidative stress causes mitochondrial depolarization leading to apoptosis-like death in Leishmania donovani promastigotes. J Biol Chem 2002; 277:24717-24722.

15. Madeiro RF, Benchimol B. The effect of drugs in T. foetus. Parasitol Res 2004; 92:159-170.

16. Müller M. Structure. In: Honigberg BM, ed. Trichomonads parasitic in humans. New York: Springer-Verlag, 1990:5-35.

17. Benchimol M, De Souza W. Fine structure and cytochemistry of the hydrogenosome of Tritrichomonas foetus. J Protozool 1983; 30:422-425.

18. Benchimol M, Johnson PJ, de Souza W. Morphogenesis of the hydrogenosome: an ultrastructural study. Biol Cell 1996; 87:197-205.

19. Johnson PJ, Lahti CJ, Bradley PJ. Biogenesis of the hydrogenosome in the anaerobic protist Trichomonas vaginalis. J Parasitol 1993; 79:664-670.

20. Rosa IA, Einicker-Lamas M, Bernardo RR et al. Cardiolipin in Hydrogenosomes: Evidence of Symbiotic Origin. Eukaryot Cell 2006; 5:784-787.

21. Lindmark DG, Müller M. Hydrogenosome, a cytoplasmic organelle of the anaerobic flagellate Tritrichomonas foetus and its role in pyruvate metabolism. J Biol Chem 1973; 248:7724-7728.

22. Clemens DL, Johnson PJ. Failure to detect DNA in hydrogenosomes of Trichomonas vaginalis by nick translation and immunomicroscopy. Mol Biochem Parasitol 2000; 106:307-313.

23. Embley TM, Horner DA, Hirt RP. Anaerobic eukaryote evolution: hydrogenosomes as biochemically modified mitochondria. TREE 1997; 12:433-441.

24. Dyall SD, Koehler CM, Delgadillo-Correa MG et al. Presence of a member of the mitochondrial carrier family in hydrogenosomes: Conservation of membrane-targeting pathways between hydrogenosomes and mitochondria. Mol Cell Biol 2000; 20:2488-2497.

25. Arnoult D, Tatischeff I, Estaquier J et al. On the evolutionary conservation of the cell death pathway: mitochondrial release of an apoptosis-inducing factor during Dictyostelium discoideum cell death. Mol Biol Cell 2001; 12:3016-3030.

26. Arnoult D, Akarid K, Grodet A et al. On the evolution of programmed cell death: apoptosis of the unicellular eukaryote Leishmania major involves cysteine proteinase activation and mitochondrion permeabilization. Cell Death Differ 2002; 9:65-81

27. Lee N, Bertholet S, Debrabant A et al. Programmed cell death in the unicellular protozoan parasite Leishmania. Cell Death Differ 2002; 9:53-64.

28. Sen N, Das BB, Gaguly A et al. Camptothecin-induced imbalance in intracellular cation homeostasis regulates programmed cell death in unicellular hemoflagellate Leishmania donovani. J Biol Chem 2004; 279:52366-52375.

29. Brenner C, Kroemer G. Apoptosis. Mitochondria: the death signal integrators. Science 2000; 289:1150-1151.

30. Kroemer G, Reed JC. Mitochondrial control of cell death. Nat Med 2000; 6:513-519.

31. Green DR, Reed JC. Mitochondria and apoptosis. Science 1998; 281:1309-1312.

32. Loeffler M, Kroemer G. The mitochondrion in cell death control: certainties and incognita. Exp Cell Res 2000; 256:19-26.

33. Ferri KF, Kroemer GK. Organelle-specific initiation of cell death pathways. Nat Cell Biol 2001; 3: E255-E263.

34. Lindmark DG, Müller M. Superoxide dismutase in the anaerobic flagellates, Tritrichomonas foetus and Monocercomonas sp. J Biol Chem 1974; 249:4634-4637.

35. Page-Sharp M, Behm CA, Smith GD. Tritrichomonas foetus and Trichomonas vaginalis: the pattern of inactivation of hydrogenase activity by oxygen and activities of catalase and ascorbate peroxidase. Microbiology 1996; 142:207-211.

36. Ryley JF. Studies on the metabolism of the protozoa. 5. Metabolism of the parasitic flagellate Tritrichomonas foetus. Biochem J 1955; 59:361-369.
37. Müller M. Biochemical cytology of trichomonad flagellates. I. Subcellular localization of hydrolases, dehydrogenases and catalase in Tritrichomonas foetus. J Cell Biol 1973; 57:453-474.
38. Benchimol M, Almeida JCA, de Souza W. Further studies on the organization of the hydrogenosome in Tritrichomonas foetus. Tissue Cell 1996; 28:287-299.
39. Sesso A, Fujiwara DT, Jaeger M et al. Structural elements common to mitosis and apoptosis. Tissue Cell 1999; 31:357-371.
40. Mancini M, Machamer CE, Roy S et al. Caspase-2 is localized at the Golgi complex and cleaves golgin-160 during apoptosis. J Cell Biol 2000; 149:603-612.
41. Lane JD, Lucocq J, Pryde J et al. Caspase-mediated cleavage of the stacking protein GRASP65 is required for Golgi fragmentation during apoptosis. J Cell Biol 2002; 156:495-509.
42. Chiu R, Novikov L, Mukherjee S et al. A caspase cleavage fragment of p115 induces fragmentation of the Golgi apparatus and apoptosis. J Cell Biol 2002; 159:637-648.
43. Ribeiro KC, Monteiro-Leal LH, Benchimol M. Contributions of the axostyle and flagella to the closed mitosis of Tritrichomonas foetus and Trichomonas vaginalis. J Euk Microbiol 2000; 47:481-492.
44. Ribeiro KC, Vetö Arnholdt AC, Benchimol M. Tritrichomonas foetus: induced synchrony by hydroxyurea. Parasitol Res 2002; 88:627-631.
45. Edinger AL, Thompson CB Death by design: apoptosis, necrosis and autophagy. Cur Opin Cell Biol 2004; 16:663-669.
46. Kitanaka C, Kuchino Y. Caspase-independent programmed cell death with necrotic morphology. Cell Death Differ 1999; 6:508-515.
47. Lockshin RA, Zakeri Z. Apoptosis, autophagy and more. Int J Biochem Cell Biol 2004; 36: 2405-2419.
48. Benchimol M, Almeida AJC, Lins U et al. Electron microscopy study of the effect of zinc in Tritrichomonas foetus. Antimicrob Agents Chemoth 1993; 2722-2726.
49. Sperandio S, de Belle I, Bredesen DE. An alternative non-apoptotic form of programmed cell death. Proc Natl Acad Sci 2000; 96:14376-14391.

CHAPTER 10

Programmed Cell Death and the Enteric Protozoan Parasite *Blastocystis hominis:* Perspectives and Prospects

Kevin S.W. Tan*

Abstract

The propensity for unicellular eukaryotes to undergo programmed cell death (PCD) has been well documented in recent years. This fascinating yet somewhat counterintuitive phenomenon has been reported to occur for many species of the parasitic Protozoa. Among the luminal Protozoa, PCD in *Blastocystis* has been the best characterized. This intestinal protozoan parasite has been shown to exhibit a number of PCD features that are apoptotic or non-apoptotic. Caspase-like activity and mitochondria are involved in *Blastocystis* PCD and have been linked to DNA fragmentation in the parasite. PCD, however, can also occur in the absence of mitochondrial and caspase involvement and DNA fragmentation. This indicates that multiple cell death pathways exist in *Blastocystis*, highlighting the cellular complexity of the seemingly simple Protozoa. Despite advances in our understanding of *Blastocystis* PCD, specific genes and proteins associated with this process have yet to be identified. Recent work has shown that *Blastocystis* can induce apoptosis in intestinal epithelial cells in vitro. Interplay of cell death between host and parasite should provide interesting and novel insights into host-pathogen interactions in vivo.

Introduction

Programmed cell death (PCD) is an important feature of multicellular organisms and functions in development, regulation of cell numbers, response to external stimuli and eradication of intracellular pathogens. There has been tremendous interest in the scientific community to characterize and understand the molecular mechanisms of PCD, primarily because of its roles in health and disease. While PCD has been best characterized among members of the Metazoa, the study of PCD in the Protozoa is a much more recent event. This delay was attributed to the belief that unicellular organisms possessed no need for a cell death programme, since death of the cell meant the demise of the entire organism.[1] However there has been numerous recent reports of PCD features in unicellular organisms including bacteria, yeast, trypanosomes, *Leishmania, Tetrahymena, Dictyostelium* (as reviewed in ref. 1) and *Blastocystis*.[2] It has been proposed that unicellular organisms do not live in isolation but interact as complex communities very much like their multicellular counterparts.[3] The activation of a PCD pathway may benefit unicellular organisms in a variety of ways. The controlled dismantling and packaging of apoptotic cells into smaller membrane-bound apoptotic bodies as described for *B. hominis* may serve to prevent exposure of cytotoxic contents of dying cells to healthy neighboring ones. It has also been suggested that a fraction of the insect vector

*Kevin S.W. Tan—Laboratory of Molecular and Cellular Parasitology, Department of Microbiology, Yong Loo Lin School of Medicine National University of Singapore, 5 Science Drive 2, Singapore 117597. Email: mictank@nus.edu.sg

Programmed Cell Death in Protozoa, edited by José Manuel Pérez Martín.

procyclic form of *T. brucei* undergoes apoptosis in hosts so as to maintain stable cell densities and to avoid competition for nutrients.[4] More recently, it has been observed that interactions between parasites and host cells mimic in vivo PCD processes in the host, such as exposure of *Leishmania* phosphatidylserine (PS) residues to host macrophages, resulting in down regulation of the host immune response and persistence of the intracellular state (Wanderley et al 2006).[5] Despite accumulating reports describing PCD in unicellular eukaryotes, the field of protozoan PCD is still in its infancy. Nonetheless, the identification of parasite-specific molecules and pathways associated with PCD opens new and exciting possibilities for antiparasitic therapies.

Blastocystis is an enigmatic protozoan of the intestinal tract. It is a polymorphic organism for which the vacuolar, granular, amoeboid and cystic forms are commonly described.[6] The clinical relevance of *Blastocystis* is currently controversial as there are numerous reports that either implicate of exonerate the protozoan as a cause of intestinal disease (as reviewed in ref. 6), although recent work from our lab[7] has shown that *Blastocystis* is able to induce contact-dependent apoptosis, F-actin rearrangement and barrier function disruption in intestinal epithelial cells (IEC). The parasite is transmitted via the fecal-oral route and is prevalent in regions with poor hygiene and deficient sanitation facilities. There are a number of fascinating aspects to *Blastocystis* biology, including extensive yet cryptic genetic diversity, the presence of unusual mitochondria-like structures and a large central vacuole (CV) with no obvious function. *Blastocystis* can also undergo PCD, in response to a variety of stimuli. This chapter aims to review *Blastocystis* PCD and highlight recent observations that suggest multiple PCD pathways exist in this organism. The possible interplay between host and *Blastocystis* PCD in relation to parasite persistence in the gut is also discussed.

PCD in *Blastocystis*—Apoptotic and Non-Apoptotic Features

PCD can be defined as a sequence of events based on cellular metabolism that leads to cell destruction.[8] The most well studied phenotype of PCD is apoptosis, which is characterized by cell shrinkage, maintenance of plasma membrane integrity, DNA fragmentation and formation of apoptotic bodies, followed by ordered removal via phagocytosis.[8] Apoptosis occurs upon intrinsic (DNA damage, viral infection) or extrinsic (FasL, serum starvation) stimuli and involves the activation of a family of cysteine proteases called caspases. Cellular signaling processes involving caspases ultimately lead to the characteristic phenotypes seen in apoptotic death. In recent years there have been descriptions of PCD occurring in the absence of caspases. These caspase-independent forms of cell death have unique morphological and biochemical characteristics and may be classified as programmed necrosis, paraptosis or autophagy.[9,10] Apoptotic features were described in *Blastocystis* cells exposed to a surface-reactive cytotoxic antibody[11] and the drug metronidazole.[12] In vitro cultures of *Blastocystis* cells are highly amenable to flow cytometric analysis as these generally comprise of discrete spherical cells that are between 5–30 μm in diameter. The use of flow cytometry, once restricted to the study of cells of the immune system, has also been shown to be useful for the analysis of Protozoa.[13] The flow cytometer records deflection patterns of laser beams after they hit individual cells flowing in a steady stream of carrier fluid. Various properties such as DNA content, cell size, cell granularity and fluorescence intensity (if fluorescent probes are used) can be obtained rapidly and analysed objectively for large numbers of cells. This technique has been enormously useful in PCD studies of *Blastocystis*. Perhaps the most dramatic phenotype observable upon PCD induction of *Blastocystis* is the reduction in cell size, evidenced by a reduction in forward scatter in flow cytometry histograms. The phenomenon is termed apoptotic volume decrease (AVD) and, in mammalian cells, is an early event linked to the efflux of intracellular ions out of the cell.[14] Little is known about AVD in *Blastocystis* but this is certainly an interesting area for further investigations since AVD is of emerging interest in the PCD community. *Blastocystis* cells exposed to both cytotoxic drug and specific antibody retained plasma membrane integrity during initial stages of cell death, evidenced by fluorescein diacetate labeling and propidium iodide exclusion.[11,12] This fulfilled the criteria for retention of membrane integrity as an apoptotic feature, in contrast to necrosis, where the plasma membrane is compromised. Annexin V-FITC assays clearly showed that apoptotic *Blastocystis* cells displayed externalization of plasma membrane PS residues as early as two hours post-induction, another typical feature of apoptosis.[11,12] In mammalian apoptosis, externalization of PS signals neighboring cells to phagocytose apoptotic cells and/or apoptotic bodies to minimize the inflammatory response.[15] The role of PS externalization in apoptotic *Blastocystis* cells has yet to be determined, although it is possible that it serves a similar function in the clearance of apoptotic bodies (see section Apoptosis of Host Cells by *Blastocystis*).

Transmission electron micrographs revealed a number of ultrastructural features suggestive of apoptosis. This included smaller electron-dense cells, cytoplasmic vacuoles and margination of nuclear chromatin.[11,12,16] Ultrastructural observations on antibody and drug exposed cells also revealed the formation of cytoplasmic blebbing and apoptotic bodies, key features in apoptosis. In the Metazoa, these membrane-bound structures, usually containing organelles or remnants thereof, bud off from the cell plasma membrane. Interestingly, apoptotic *Blastocystis* cells appeared to deposit apoptotic bodies into the large central vacuolar space by an invagination process. This finding implicates a novel role for the *Blastocystis* CV; it serves as a repository for apoptotic bodies during PCD, delaying their release into the extracellular environment. The postulated sequence of events that occur during *Blastocystis* apoptotic PCD is represented in Figure 1. Another form of PCD is autophagic cell death, in which dying cells deposit cytoplasmic contents via specialized structures called autophagic vacuoles, which are subsequently targeted to lysosomes.[8,9] We had previously observed autophagic-like features in aging cells grown as colonies (see section Alternative Deathstyles of *Blastocystis*). These colony forms also displayed degenerating organelles within the CV. Perhaps the CV also acts as a large autophagic vacuole, a site for degradation and recycling of

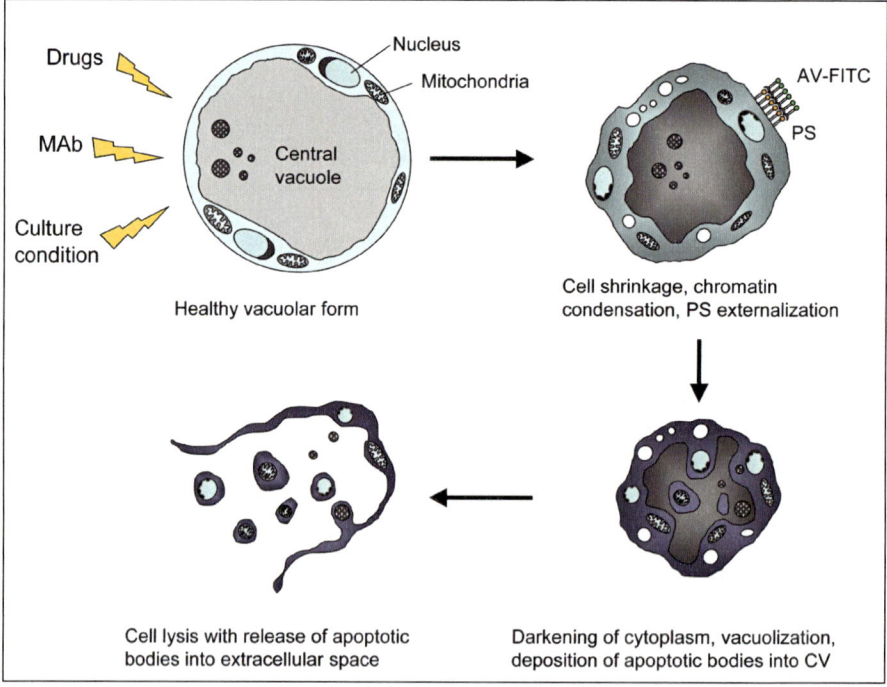

Figure 1. Postulated model for *Blastocystis* apoptotic cell death. Apoptotic features are observed upon exposure to the drug metronidazole,[12] cytotoxic monoclonal antibody[11] and after colony growth in vitro.[16] Early features include cell shrinkage and externalization of phosphatidylserine (PS) residues, detectable by annexin V-FITC (AV-FITC). At later stages, vacuolization and darkening of cytoplasm are evident, concomitant with the deposition of apoptotic bodies into the central vacuole (CV). Apoptotic bodies are eventually released into the extracellular space. (Figure adapted from ref. 11.)

cytoplasmic material during PCD or periods of cellular starvation. More studies involving specific markers for and inhibitors of autophagic vacuoles are required to validate this hypothesis.

Apoptotic *Blastocystis* cells undergo extensive in situ DNA fragmentation but did not reveal typical DNA laddering patterns on agarose gels despite clear evidence for in situ DNA[11,12] and nuclear fragmentation. The authors suggested that this inconsistency could be due to asynchronous DNA fragmentation, lack of sensitivity of the gel electrophoresis approach, or the formation of large DNA fragments not resolvable in a 2% agarose gel. It has been reported that DNA degradation in the Metazoa occurs in a two stage process; large fragments of 50-300 kb are initially generated via cleavage of the nuclear scaffold, followed by cleavage of nucleosomal spacer regions to generate the classical 180 bp ladder.[17] It has been suggested that each stage is contributed by different molecular activities, with the generation of large fragments playing a greater role in apoptosis than the formation of oligonucleosomes.[18] Whether this is also the case for *Blastocystis* PCD warrants attention and should be easily addressed with the use of pulsed-field gel electrophoresis to better visualize fragmentation of high molecular weight DNA in dying parasites.

Caspase and Mitochondrial Involvement

Caspases, key components of the apoptotic death signaling pathway are synthesized as inactive precursors that are activated upon proteolytic cleavage. They can be broadly classified as initiator and effector caspases. Initiator caspases (caspase-2, -8, -9 and -10) are characterized by longer N-terminal prodomains and function to activate the downstream effector caspases (caspase-3, -6 and -7).[19] Caspase-3 is widely involved in the execution of apoptosis and its proteolytic cleavage of subcellular substrates results in the features seen in this form of cell death, including DNA fragmentation. Apoptotic stimuli may also induce the dysregulation of mitochondrial membrane potential ($\Delta\Psi_m$), resulting in the release of a variety of pro-apoptotic molecules such as cytochrome c and endonuclease G. DNA fragmentation and cell death can be mediated via the mitochondrial pathway, which may be caspase-dependent or –independent. We had shown evidence for caspase-3-like antigens in *B. hominis* and observed time-dependent increase in caspase-3 activity after exposure to 1D5 by colorimetric, flow cytometry and microscopy approaches.[20] This activity was abolished when cells were pretreated with the caspase-3 inhibitor Ac-DEVD-CHO. In most metazoan cell lines, inhibition of caspase-3 results in partial or complete inhibition of DNA fragmentation.[21,22] Inhibition of *Blastocystis* caspase-3 resulted in partial inhibition of DNA fragmentation[23] and did not rescue the cells from PCD, suggesting that caspase-independent apoptotic pathways might also be involved. Using the pancaspase inhibitor zVAD.fmk, DNA fragmentation was also partially inhibited, indicating that DNA fragmentation was not wholly dependent on caspases. Complete abolishment of DNA fragmentation was achieved by combined use of zVAD.fmk and cyclosporin A (CA), a mitochondrial transition pore blocker.[24] The observation that DNA fragmentation in *Blastocystis* was not completely blocked by zVAD.fmk treatment, but was abolished when the CA was added suggests that caspase-independent mitochondrial pathways exist to induce DNA breaks during *Blastocystis* PCD. In mammalian systems, mitochondrial release of endonuclease G and apoptosis-inducing factore (AIF) lead to DNA fragmentation in the absence of caspase activation.[8] The proapoptotic mitochondrial factors in *Blastocystis* have yet to be identified. Interestingly, the mitochondrial-like organelles of *Blastocystis* have been reported to lack cytochromes.[25] We had recently described a technique for the isolation and purification of mitochondrial-like organelles from *Blastocystis*[26] and this approach should be useful for the identification of novel apoptotic regulatory molecules harbored within these structures. A surprising finding was that the complete abolishment of DNA fragmentation by both inhibitors failed to rescue the cells from PCD, evidenced by reduced cell counts and AVD.[24] We conclude that other pathways must exist to activate PCD in *Blastocystis* (Fig. 2). Such caspase- and mitochondria -independent PCD has only recently been described for higher eukaryotes.[27] Taken together, *Blastocystis* cell death molecular pathways appear to be as complex as those of mammalian cells. Interestingly, our results therefore show that the specific pancaspase inhibitor, zVAD.fmk, did not rescue cells from MAb 1D5-induced AVD despite causing a reduction in DNA degradation. There are conflicting reports on the association

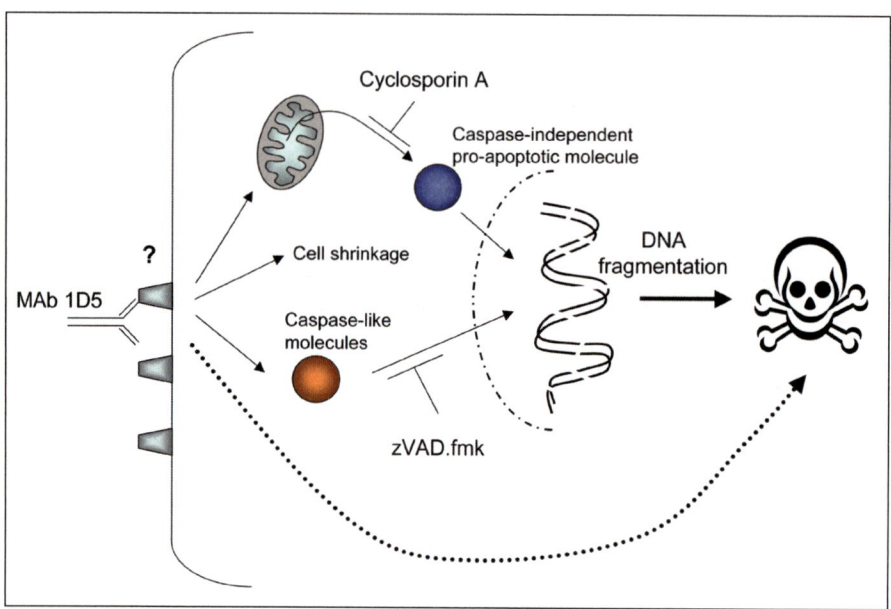

Figure 2. Proposed model of PCD in *B. hominis*. MAb 1D5 binds to an unidentified 30 kDa antigen on the surface of *B. hominis*. This binding activates caspase-like proteases and also leads to permeabilization of the outer mitochondrial membrane. The binding of MAb 1D5 also causes cell shrinkage. These events eventually lead to DNA fragmentation and cell death. However, cell death can also occur in the absence of DNA fragmentation by an as yet unidentified pathway. (Figure adapted from ref. 24.)

between cell volume decrease and DNA fragmentation in certain metazoan systems. Bortner and Cidlowski[28] reported that cell shrinkage can be separated from other characteristic features of apoptosis, with sodium ions controlling cell size and potassium ions regulating apoptosis. Vu et al[29] showed that in UV-C-induced death of Jurkat cells, the loss of cell volume varies depending on the specific apoptotic pathway that is activated. However, in UV-C-induced death of Jurkat cells, pretreatment with a specific caspase-3 inhibitor had no effect on cell shrinkage.[30] Hence our observation that AVD and DNA fragmentation are apparently uncoupled processes in *Blastocystis* is interesting and warrants more research.

Alternate Deathstyles of *Blastocystis*—More than One Way to Kill a Parasite?

It has become increasingly clear that PCD may take on non-apoptotic forms at the morphological, biochemical and molecular level.[1,8] Two recently described forms of non-apoptotic PCD are autophagy[31] and paraptosis.[10] Autophagy is ubiquitous among diverse eukaryotic cell types and has been described in mammalian cells and in *Leishmania* species.[32] It may function to aid survival during periods of nutrient deprivation and also as part of cell remodeling during growth and development. It is unlikely that the caspase- and mitochondrial-independent cell death described for 1D5-exposed *Blastocystis* occurs via autophagy as this form of PCD, although caspase-independent, generally involves mitochondrial-dysregulation and DNA fragmentation. However, we have observed ultrastructural features closely resembling autophagy in aging *Blastocystis* cells grown as colonies in agar (ref. 16 and unpublished data). These included prominent Golgi apparatus, cytoplasmic blebbing, vacuoles containing myelin- and lipid-like inclusions and pyknotic nuclei. These features were most apparent in regions deeper within the colony, where nutrients would

Table 1. Diversity of PCD features in **Blastocystis**

	Apoptosis	Autophagy	Paraptosis	Blastocystis PCD
Apoptotic bodies/ blebbing	+	Sometimes	–	+
Inward blebbing of apoptotic bodies	–	–	–	+
Cell shrinkage	+	+[a]	+[a]	+
Cytoplasmic vacuolation	–	+	+	+
Phosphatidylserine externalization	+	?[b]	+	+
Caspase activity	+	–	–	+/–[c]
In situ DNA fragmentation	+	+	–	+/–[c]
Disruption of mitochondrial outer membrane potential	+	+	?[d]	+/–[c]

[a]There are, to date, only a few reports associating autophagy[33] or paraptosis[34] with cell shrinkage. [b]There is currently no information on whether autophagy is associated with phosphatidylserine externalization. [c]+/– refers to cell death in *Blastocystis* that has been observed to occur in the presence or absence of these features. [d]There is currently no clear indication if paraptosis is associated with disruption of mitochondrial outer membrane potential. (Table modified from ref. 2.)

be limiting. This observation supports the role of autophagy in turnover of organelles during cellular starvation. Paraptosis is a non-apoptotic form of PCD that occurs in the absence of caspase activation or typical nuclear changes but not much is known about mitochondrial involvement.[10] Cell shrinkage (AVD) in *Blastocystis* undergoing PCD may not be a useful factor to discriminate between apoptosis, autophagy or paraptosis. This is because all three forms of PCD have been reported to exhibit AVD,[8,33,34] although the association between AVD and non-apoptotic PCD is currently not well established, due to limited studies. Future studies should therefore consider the possibility that the caspase- and mitochondria-independent cell death in *Blastocystis* occurs by paraptosis. Taken together, it is highly suggestive that the protozoan parasite *Blastocystis* possesses multiple mechanisms of PCD, as has recently been documented for higher eukaryotes and is in line with increasing reports, in the Metazoa, of alternative, non-apoptotic PCD existing in parallel with apoptosis. The novel molecules that regulate these pathways may provide new targets for chemotherapy especially if these molecules are conserved among the parasitic protozoa but are not found humans. Table 1 summarizes the diverse PCD features in *Blastocystis*.

Apoptosis of Host Cells by *Blastocystis*

The ability of pathogenic microorganisms to induce cell death, notably by apoptosis, in host cells has been well documented.[35-38] We recently report that *Blastocystis* induces contact-independent apoptosis in IEC-6 cell lines in vitro.[7] Classical apoptotic features such as in situ DNA fragmentation, nuclear blebbing, caspase-3 activation and externalization of PS residues were observed in host cells exposed to *Blastocystis* parasites. The existence of conserved apoptotic features in both host and pathogen provides us with novel perspectives on host-pathogen interactions and may shed new light on parasite persistence and pathogenesis. It was recently reported that *L. amazonensis* amastigotes signal host macrophages via exposed PS, which is an important mechanism for leishmanial establishment in the vertebrate host.[5] The ingested PS-exposed parasites apparently mimic host PS-exposed apoptotic bodies and results in the induction of an anti-inflammatory response.

The survival of the parasite is dependent on the amount of PS that is exposed to the macrophage, which is in turn dependent on the mouse strain used previously for infection. It was observed that macrophage leishmanicidal activity is decreased when infected with BALB/c derived amastigotes, as these cells produce more TGFβ1 and less NO than macrophages infected with C57BL/6-derived amastigotes. Such 'apoptotic mimicry' among microorganisms may be prevalent in nature and could be important for parasite survival in or on host cells.

This fascinating phenomenon may also operate in *Blastocystis*-intestinal epithelia interactions, since *Blastocystis* cells dying by PCD do externalize PS[2] and dying colony forms produce numerous apoptotic body-like particles of various sizes.[16] It has been observed that apoptotic bodies of IEC are internalized by neighbouring cells and by underlying macrophages.[39,40] Phagocytosis of PS-exposed apoptotic bodies has been shown to inhibit pro-inflammatory immune responses by macrophages.[41,42] We hypothesize that, in vivo, PS-exposed apoptotic bodies of *Blastocystis* are recognized and are ingested by local host macrophages. This serves to down regulate host pro-inflammatory responses, promoting parasite survival. The effect of *Blastocystis* on IEC turnover is another important area to study in light of recent data indicating that accelerated IEC turnover is a mechanism of parasite expulsion in the host.[43]

Implications and Future Directions

It has been suggested that PCD is an ancient mechanism of self-destruction that existed before the advent of multicellularity.[1] Hence, it is believed that all unicellular eukaryotes have the propensity to undergo PCD. *Blastocystis hominis* exhibits diverse PCD pathways that appear apoptotic, autophagic, paraptotic, or may even be completely novel. This sheds new light on the cellular complexity of the seemingly simple Protozoa. In retrospect, earlier reports of unusual cellular structures and phenomenon may be attributed to *Blastocystis* undergoing PCD. The proposal that *Blastocystic* undergoes a multiple-fission-like division, akin to apicomplexan schizogony, is highly controversial.[44-47] Proponents for this form of cell replication report progeny-like structures within the CV based predominantly on light and fluorescence microscopy.[44,45,47] However, others argue that these reports lack convincing ultrastructural and growth profile data.[46] The subcellular progeny may, in reality, be apoptotic bodies within the CV or autophagic vacuoles containing nuclear, mitochondrial or other organelle structures. Vdovenko[48] observed that the granular form is a product of rapid cell degeneration and represents a stage prior to cell death. These CV granules seen during cellular degeneration may represent the apoptotic bodies we had reported previously in cells undergoing PCD.[11,12]

The use of autophagic markers and inhibitors, such as monodansylcadaverine and 3-methyladenine respectively, would help confirm the existence of autophagy in *Blastocystis*. Despite conservation at the biochemical pathway and morphological level, there are surprisingly few genes and proteins implicated as activators or executors of protozoan PCD.[1] Despite this limitation, the availability of genome data for an increasing number of unicellular eukaryotes has greatly aided protozoan PCD studies, as this facilitates the screening for conserved gene orthogues that may be further investigated for PCD involvement.[49-51] Though currently unavailable, a *Blastocystis* genome database would greatly facilitate the screening of genes directly or indirectly involved in PCD. PCD is frequently triggered by events that alter the expression of key target genes. Such genes can be identified by techniques that analyze gene expression. Suppressive subtractive hybridization (SSH) is an established technique useful in nonstandard model organisms for which comprehensive gene microarrays are not available or, as in our case, limited homology exists between protozoan and metazoan PCD-related molecules. SSH has been successfully employed for the identification of genes involved in mammalian PCD[52] and we are currently employing this technique to identify genes involved in apoptotic and non-apoptotic *Blastocystis* PCD. Although PCD has been known to involve genes regulated at the transcriptional level, it is also possible for PCD to be regulated at the posttranscriptional or translational level.[53] In this regard, a proteomics approach to identify proteins involved in *Blastocystis* PCD would then have to be adopted.

Recently, metacaspases, a group of cysteine proteases present in unicellular eukaryotes and belonging to the caspase/paracaspase/metacaspase superfamily, has been suggested to play a role in yeast and protozoan PCD.[50,54] Future studies are required to identify and functionally characterize metacaspases in *Blastocystis*. The questionable existence of *Blastocystis* metacaspases notwithstanding, presence of caspase-like molecules and activity in *Blastocystis* and other parasites provide novel opportunities for antiparasite therapy by drugs that work as caspase activators. Pharmacological activation of caspases using small molecules, or lowering their activation threshold may trigger cell death in parasites, as has been suggested for cancer cells.[19] *Blastocystis* possesses unusual mitochondria that have been reported to lack cytochromes.[55] Interestingly, most mitochondrial-dependent apoptotic pathways involve the mitochondrial release of cytochrome c into the cytoplasm to initiate the cell death cascade. Hence, the characterization of the unusual mitochondria of *Blastocystis* may reveal novel mitochondrial mediators of PCD. The PCD studies on *Blastocystis* and other parasitic Protozoa have shown that PCD pathways in these organisms have both conserved and unique properties. The interplay between cell death processes and molecules between host and parasite should lead to new insights into parasite virulence and persistence in the host. As well as giving us new knowledge on the origin of PCD, research on the novel mediators of protozoan PCD should also lead to new targets for treatment and prevention of parasitic diseases.

Acknowledgements

Our work on this topic was supported by grants R-182–000–058–213 and R-182-000-090-305 from the National Medical Research Council and Biomedical Research Council respectively. We are grateful to AMA Nasirudeen, Manoj K Puthia, Ng Geok Choo, NP Ramachandran and Josephine Howe for constructive discussions and technical support.

References

1. Ameisen JC. On the origin, evolution and nature of programmed cell death: a timeline of four billion years. Cell Death Differ 2002; 9:367-393.
2. Tan KSW, Nasirudeen AMA. Protozoan programmed cell death: insights from Blastocystis deathstyles. Trends Parasitol 2005; 21:547-550.
3. DosReis GA, Barcinski MA. Apoptosis and parasitism: from the parasite to the host immune response. Adv Parasitol 2001; 49:133-161.
4. Welburn SC, Maudlin I. Tsetse-trypanosome interactions: rites of passage. Parasitol Today 1999; 15:399-403.
5. Wanderley JL, Moreira ME, Benjamin A et al. Mimicry of apoptotic cells by exposing phosphatidylserine participates in the establishment of amastigotes of Leishmania (L) amazonensis in mammalian hosts. J Immunol 2006; 176:1834-1839.
6. Tan KSW. Blastocystis in humans and animals: new insights using modern methodologies. Vet Parasitol 2004; 26:121-144.
7. Puthia MK, Sio SWS, Jia L et al. Blastocystis induces contact-independent apoptosis, F-actin rearrangement and barrier function disruption in IEC-6 cells. Infect Immun 2006; 74:4114-4123.
8. Guimarães CA, Linden R. Programmed cell death: apoptosis and alternative deathstyles. Eur J Biochem 2004; 271:1638-1650.
9. Edinger AL, Thompson CB. Death by design: apoptosis, necrosis and autophagy. Curr Opin Cell Biol 2004; 16:663-669.
10. Sperandio S, de Belle I, Bredesen DE. An alternative, nonapoptotic form of programmed cell death. Proc Natl Acad Sci USA 2000; 97:14376-14381.
11. Nasirudeen AMA, Tan KSW, Singh M et al. Programmed cell death in a human intestinal parasite, Blastocystis hominis. Parasitology 2001; 123:235-246.
12. Nasirudeen AMA, Yap EH, Singh M et al. Metronidazole induces programmed cell death in the protozoan parasite Blastocystis hominis. Microbiology 2004; 150:33-43.
13. Vesey G, Hutton P, Champion A et al. Application of flow cytometric methods for the routine detection of Cryptosporidium and Giardia in water. Cytometry 1994; 16:1-6.
14. Bortner CD, Cidlowski JA. Apoptotic volume decrease and the incredible shrinking cell. Cell Death Differ 2002; 9:1307-1310.
15. Adayev T, Estephan R, Meserole S et al. Externalization of phosphatidylserine may not be an early signal of apoptosis in neuronal cells, but only the phosphatidylserine-displaying apoptotic cells are phagocytosed by microglia. J Neurochem 1998; 71:1854-1864.

16. Tan KSW, Howe J, Yap EH et al. Do Blastocystis colony forms undergo programmed cell death? Parasitol Res 2001; 87:362-367.
17. Nagata S, Nagase H, Kawane K et al. Degradation of chromosomal DNA during apoptosis. Cell Death Differ 2003; 10:108-116.
18. Walker PR, Sikorska M. New aspects of the mechanism of DNA fragmentation in apoptosis. Biochem Cell Biol 1997; 75:287-299.
19. Philchenkov A. Caspases: potential targets for regulating cell death. J Cell Mol Med 2004; 8:432-444.
20. Nasirudeen AMA, Singh M, Yap EH et al. Blastocystis hominis: evidence for caspase-3-like activity in cells undergoing programmed cell death. Parasitol Res 2001; 87:559-565.
21. Del Bello B, Valentini MA, Mangiavacchi P et al. Role of caspases-3 and -7 in Apaf-1 proteolytic cleavage and degradation events during cisplatin-induced apoptosis in melanoma cells. Exp Cell Res 2004; 293:302-310.
22. Janicke RU, Sprengart ML, Wati MR et al. Caspase-3 is required for DNA fragmentation and morphological changes associated with apoptosis. J Biol Chem 1998; 273:9357-9360.
23. Nasirudeen AMA, Tan KSW. Caspase-3-like protease influences but is not essential for DNA fragmentation in Blastocystis undergoing apoptosis. Eur J Cell Biol 2004; 83:477-482.
24. Nasirudeen AMA, Tan KSW. Programmed cell death in Blastocystis hominis occurs independently of caspase and mitochondrial pathways. Biochimie 2005; 87:489-497.
25. Nakamura Y, Hashimoto T, Yoshikawa H et al. Phylogenetic position of Blastocystis hominis that contains cytochrome-free mitochondria, inferred from the protein phylogeny of elongation factor 1 alpha. Mol Biochem Parasitol 1996; 77:241-245.
26. Nasirudeen AMA, Tan KSW. Isolation and characterization of the mitochondrion-like organelle from Blastocystis hominis. J Microbiol Methods 2004; 58:101-109.
27. Godefroy N, Lemaire C, Renaud F et al. p53 can promote mitochondria- and caspase-independent apoptosis. Cell Death Diff 2004; 11:785-787.
28. Bortner CD, Cidlowski JA. Uncoupling cell shrinkage from apoptosis reveals that Na+ influx is required for volume loss during programmed cell death. J Biol Chem 2003; 278:39176-39184.
29. Vu CC, Bortner CD, Cidlowski JA. Differential involvement of initiator caspases in apoptotic volume decrease and potassium efflux during Fas-and UV-induced cell death. J Biol Chem 2001; 276:37602-37611.
30. Scoltock AB, Cidlowski JA. Activation of intrinsic and extrinsic pathways in apoptotic signaling during UV-C-induced death of Jurkat cells: the role of caspase inhibition. Exp Cell Res 2004; 297:212-223.
31. Klionsky DJ, Emr SD. Autophagy as a regulated pathway of cellular degradation. Science 2000; 290:1717-1721.
32. Bera A, Singh S, Nagaraj R et al. Induction of autophagic cell death in Leishmania donovani by antimicrobial peptides. Mol Biochem Parasitol 2003; 127:23-35.
33. von Bultzingslowen I, Jontell M, Hurst P et al. 5-Fluorouracil induces autophagic degeneration in rat oral keratinocytes. Oral Oncol 2001; 37:537-544.
34. Schneider D, Gerhardt E, Bock J et al. Intracellular acidification by inhibition of the Na+/H+-exchanger leads to caspase-independent death of cerebellar granule neurons resembling paraptosis. Cell Death Differ 2004; 11:760-770.
35. Chin AC, Teoh DA, Scott KGE et al. Strain dependent induction of enterocyte apoptosis by Giardia lamblia disrupts epithelial barrier function in a caspase-3-dependent manner. Infect Immun 2002; 70:3673-3680.
36. Fiorentini C, Fabbri A, Falzano L et al. Clostridium difficile toxin B induces apoptosis in intestinal cultured cells. Infect Immun 1998; 66:2660-2665.
37. Huston CD, Houpt ER, Mann BJ et al. Caspase 3-dependent killing of host cells by the parasite Entamoeba histolytica. Cell Microbiol 2000; 2:617-625.
38. Zychlinsky A, Sansonetti P. Perspectives series: host/pathogen interactions. Apoptosis in bacterial pathogenesis. J Clin Invest 1997; 100:493-495.
39. Groos S, Busche R, von Engelhardt W et al. Excessive apoptosis of guinea pig colonocytes may lead to an imbalance between phagocytosis and degradation in vivo. Cell Tissue Res 2004; 316:77-86.
40. Hall PA, Coates PJ, Ansari B et al. Regulation of cell number in the mammalian gastrointestinal tract: the importance of apoptosis. J Cell Sci 1994; 107:3569-3577.
41. Fadok VA, Bratton DL, Konowal A et al. Macrophages that have ingested apoptotic cells in vitro inhibit proinflammatory cytokine production through autocrine/paracrine mechanisms involving TGF-beta, PGE2 and PAF. J Clin Invest 1998; 101:890-898.
42. Hoffmann PR, Kench JA, Vondracek A et al. Interaction between phosphatidylserine and the phosphatidylserine receptor inhibits immune responses in vivo. J Immunol 2005; 174:1393-1404.

43. Cliffe LJ, Humphreys NE, Lane TE et al. Accelerated intestinal epithelial cell turnover: a new mechanism of parasite expulsion. Science 2005; 308:1463-1465.
44. Govind SK, Khairul AA, Smith HV. Multiple reproductive processes in Blastocystis. Trends Parasitol 2002; 18:528.
45. Singh M, Suresh K, Ho LC et al. Elucidation of the life cycle of the intestinal protozoan Blastocystis hominis. Parasitol Res 1995; 81:446-450.
46. Tan KSW, Stenzel DJ. Multiple reproductive processes in Blastocystis: proceed with caution. Trends Parasitol 2003; 19:290-291.
47. Zierdt CH. Blastocystis hominis—past and future. Clin Microbiol Rev 1991; 4:61-79.
48. Vdovenko AA. Blastocystis hominis: origin and significance of vacuolar and granular forms. Parasitol Res 2000; 86:8-10.
49. Deponte M, Becker K. Plasmodium falciparum—do killers commit suicide? Trends Parasitol 2004; 20:165-169.
50. Kosec G, Alvarez VE, Aguero F et al. Metacaspases of Trypanosoma cruzi: possible candidates for programmed cell death mediators. Mol Biochem Parasitol 2006; 145:18-28.
51. Uren AG, O'Rourke K, Aravind LA et al. Identification of paracaspases and metacaspases: two ancient families of caspase-like proteins, one of which plays a key role in MALT lymphoma. Mol Cell 2000; 6:961-967.
52. Fellenberg J, Dechant MJ, Ewerbeck V et al. Identification of drug-regulated genes in osteosarcoma cells. Int J Cancer 2003; 105:636-643.
53. Harada H, Grant S. Apoptosis regulators. Rev Clin Exp Hematol 2003; 7:117-138.
54. Mottram JC, Helms MJ, Coombs GH et al. Clan CD cysteine peptidases of parasitic protozoa. Trends Parasitol 2003; 19:182-187.
55. Zierdt CH, Donnolley CT, Muller J et al. Biochemical and ultrastructural study of Blastocystis hominis. J Clin Microbiol 1988; 26:965-970.

CHAPTER 11

Programmed Cell Death in Dinoflagellates

María Segovia*

Abstract

Dinoflagellates are unicellular flagellated eukaryotes exploiting different nutritional modes although approximately half of them are photosynthetic. They are a monophyletic group, included in the lineage Alveolates. Dinoflagellates are ecologically important as components of the phytoplankton and contribute significantly to CO_2 fixation and primary productivity in the oceans. As well, they can form blooms (densities of more than a million cells per millilitre) producing red and brown tides. Recently the possible role of programmed cell death (PCD) in phytoplankton has received much attention because massive cell disappearance of species as a consequence of cell death have important consequences in the ocean dynamics. Several species of phytoplankters undergo PCD, apparently using the same core mechanism as metazoans. However, dinoflagellates show different PCD morphologies (apoptotic, necrotic, necrotic-like and paraptotic) depending on the species and on the triggering factor, probably due to their mesokaryotic condition. Similarly to metazoans, PCD in this group goes through intermediates such as ROS, and the proteins in charge of executing the cell are metacaspases (cysteinyl aspartate proteases from the caspase family, found in yeast, plants, protists and some bacteria). The acquisition of the PCD genes in these organisms goes back to ancient times when the endosymbiotic events took place. However, the intriguing point is the gene persistence through evolution, given that such genes provide the cell with negative selective pressure.

Introduction

Dinoflagellates are an extraordinary enigmatic group of nutritionally versatile single-celled protists (autotrophic, heterotrophic, parasitic, mixotrophic, and symbiotic forms -although approximately half of them are photosynthetic-).[1] At least 4000 species of dinoflagellates have been described of which many are marine and some other live in fresh waters.[2] Typically, they are considered unicellular flagellated eukaryotes. Botanists used to group them together with "algae" and zoologists with "protozoa", but to locate the dinoflagellates within a taxonomic group it has always been a difficult task. In recent years, they have been included in the non monophyletic "Kingdom Protista" and in particular within the lineage Alveolates. Alveolates also comprise some of the most familiar and numerous protist groups, including the ciliates and the apicomplexans, a group of parasitic and disease-causing protists.[3] The dinoflagellates have been reclassified in the phylum Pyrrophyta (or Dynophyta), and phylogenetic trees based on SSU rRNA, and LSU rRNA alignments from surveyed sequences of dinoflagellates available in the GenBank indicate that they have been found to be a monophyletic group,[4] i.e., they share a common ancestor.

*María Segovia—Departamento de Ecología, Facultad de Ciencias-Universidad de Málaga, Campus Universitario de Teatinos s/n, 29071-Málaga, Spain. Email: segovia@uma.es

Programmed Cell Death in Protozoa, edited by José Manuel Pérez Martín.
©2008 Landes Bioscience and Springer Science+Business Media.

Their chromosomal structure is unique in all of eukaryotic life. They have a peculiar form of nucleus, in which the chromosomes are quasi-permanently condensed throughout the cell cycle and the DNA exists in a crystal liquid state.[5] The distinctive feature is that they do not have nucleosomes and present extra nuclear spindles.[6] Until not long ago, it was thought that they also lacked histones to package their DNA. Recently, some histone-like chromosomal proteins have been detected, cloned and sequenced in several species of dinoflagellates.[7-11] The nucleus of the dinoflagellates was once considered an intermediate between the nucleoid region of prokaryotes and the true nucleus of eukaryotes, and was termed *mesokaryotic*.[12] Later studies demonstrated their eukaryotic nature and DNA sequence studies propose that dinoflagellates are true eukaryotes.[13] In addition, the discovery of plastids in apicomplexans have led to suggest that they were inherited from an alveolate ancestor common to the two groups,[14] but none of the more basal lines have them. However, nuclear genomes of dinoflagellates house heterologous genes and particularly proteobacterial genes.[15] Therefore, dinoflagellates are considered by some as truly mesokaryotes in the sense that they allow themselves the use of both systems in the same cell, and precisely this mesokaryotic nature is the responsible for their biochemical diversity.[14] How dinoflagellates are able to maintain such a peculiar nuclear structure remains a mystery.

Dinoflagellates exhibit substantial morphological variations, but the most characteristic feature is the presence of two flagella inserted into their cell wall (one is transversal and the other longitudinal). Another distinctive feature of their cell structure is the presence of a layer of vesicles (alveoli) underlying the cell membrane. In many dinoflagellates, these alveoli contain cellulose or polysaccharide microfibrils, which form a protective system of abutting *plates* joined through junctions called *sutures*. When this occurs, the cell wall is referred to as a *theca*. Dinoflagellates possessing a theca are often referred to as "*armoured dinoflagellates*" and those naked are referred to as "*unarmoured dinoflagellates*" (Fig. 1).

This group of protists makes a significant contribution to CO_2 fixation and primary productivity in the oceans.[16] Photosynthetic dinoflagellates contain chlorophyll a and c, in addition to xanthophylls and peridinin. When they live as symbionts with other organisms, they lack their typical plates and flagella, appearing spherical. The xanthophylls confer the golden-brown color typical of many photosynthetic types and gives rise to the name *zooxanthellae*, used to describe the dinoflagellates that form symbioses in invertebrates such as corals, anemones and jellyfish. Photosynthetic zooxanthellae provide their hosts with photosynthetic products and they receive nutrients from the host. Precisely, is the photosynthesis carried out by the dinoflagellate symbionts that make coral reefs one of the most productive ecosystems on Earth.

Dinoflagellates form blooms (concentrations of more than a million cells per millilitre- they reproduce primarily through asexual cell division, albeit sexual reproduction takes place when the conditions become stressful and some species can form resting cysts-) producing red and brown tides. Some of these tides are deleterious and may led to the accumulation of toxins in shellfish and fish, causing illness and even death in marine animals and humans.[17-19] These blooms have also important consequences for the trophic web, biogeochemical cycling and nutrient availability. Specially, because they are related to other oceanographic events that have consequences on biodiversity, succession, and production. Although the processes of dinoflagellate bloom decline and cell death are little understood[20,21] it has been observed that bloom conditions can cause cell death. Vardi et al[22] reported for the first time that a dinoflagellate bloom disappearance was related to programmed cell death. They observed that a bloom of the freshwater dinoflagellate *Peridinium gatunense* terminated suddenly because of increasing the pH of the water. Such an increase drastically depleted the concentration of dissolved CO_2 and stimulated the formation of reactive oxygen species (ROS) to a level that induced programmed cell death, with the participation of cysteine proteases, resembling apoptosis in animal and plant cells.

The possible role of programmed cell death (PCD) not only in dinoflagellates[22-24] but also in other unicellular organisms, has received much attention recently.[25-31] The occurrence of apoptosis in unicells is confusing because, unlike multicellular organisms, it results in complete loss of the organism and must therefore be maladaptive. However, there are many interpretations

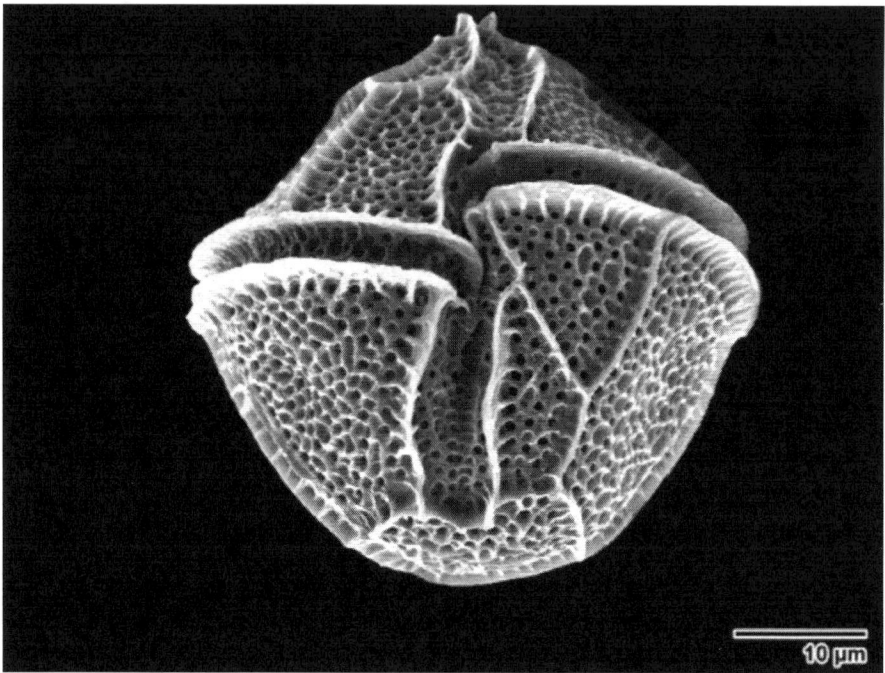

Figure 1. Scanning electron micrograph (SEM) of the marine, armoured dinoflagellate *Lingulodinium polyedrum*. This species from warm-waters is a red tide former and toxin producer. Cells of *L. polyedrum* are angular, roughly pentagonal and polyhedral-shaped. They range in size from 40-54 μm in length and 37-53 μm in transdiameter width. Thecal plates are thick, well defined, and coarsely areolate. Distinct ridges are present along the plate sutures. The SEM micrograph is courtesy of Dr. Maria Faust at the National Museum of Natural History, Smithsonian Institution, Washington DC, USA.

to its occurrence. The presence of important components of cell death pathways in some of the earliest-evolved organisms[32] suggest that their origins are truly ancient, and it has been speculated that they may be the result of viral-eukaryote genomic mixing during ancient evolutionary history.[27,31,33] Independently of its origins or evolutionary meaning, the existence of programmed cell death in dinoflagellates has important implications in aquatic ecosystems. Is PCD the main responsible of the decline of dinoflagellates blooms? Is the bleaching of zooxanthallae driven by PCD? Is genetic fitness maintained through this process or is population growth controlled by this means? Is PCD the response to abiotic stress or a consequence of viral infection?

Which Cell Death Morphology: Apoptotic, Necrotic, "Necrotic-Like" or Paraptotic?

When programmed cell death meets very specific morphological and biochemical requirements such as chromatin condensation and margination, ordered DNA cleavage while the cytoplasm and organelles remain unchanged, and the participation of caspases, it is known as apoptosis. This is clearly different to the necrotic event (loss of membrane integrity, cell swelling and lysis). However, the classical necrotic nature described as passive, unprogrammed cell death, can be questioned. There are examples of cells presenting necrotic morphologies that are subjected to an active cell death programme called "necrotic-like" PCD and the only difference with the classical definition of PCD relies on this event being caspase-independent.[34] Consistently, it is considered a type of

PCD due to the presence of underlying regulatory mechanisms. Because of the particular nature of dinoflagellates, some of them have shown apoptotic-like cell death, whilst others present characteristics difficult to appoint to a particular well defined type of PCD. Nevertheless, we find that although some groups undergo nonnecrotic cell death and share similar aspects to the apoptotic event, they are in no-man's land, showing a mixture of hallmarks typically found in both apoptotic and necrotic definitions of cell death. The explanation for this may come from the ways through which cell death has evolved from ancient times, where only prokaryotic and unicellular eukaryotes forms of live existed until multicellular organisms appeared. The mesokaryotic condition of dinoflagellates, acquiring prokaryote genes during evolution, may explain the peculiarity of the nature of their cell death programmes. Recently, a novel type of programmed cell death, which does not fulfill completely either the typical apoptotic features or the necrotic ones, has been reported as paraptosis.[35] The features of paraptosis differ from those of apoptosis and involve cytoplasmic vacuolization, mitochondrial swelling and absence of caspase activation or typical nuclear changes, including pyknosis and DNA fragmentation.[35,36] The evidence that this alternative, nonapoptotic PCD exists in parallel with apoptosis has important implications for understanding the type of PCD that occurs in unicellular organisms. Paraptosis is a form of PCD that might be particularly relevant for prokaryotic and eukaryotic unicellular microorganisms with ancient origins, such as those observed in phytoplankton.

To illustrate this, I will take the example of zooxanthellae. *Symbiodinium sp.* is the symbiotic zooxanthellae of the sea anemone *Aiptasia sp.* It is an example of species undergoing PCD and displaying apoptotic and necrotic morphologies when subjected to experimental bleaching (see the "coral bleaching" section later on in this chapter for detailed information). It was observed that when exposed to hyperthermic treatment (33-34°C) from 0 to 7 days, both necrosis and PCD occurred.[23] The release of zooxanthellae from the host—that is the very "coral bleaching" phenomenon—was associated with widespread necrosis of the animal endoderm cells. After only 4 days of hyperthermia, they showed that the onset of necrosis had taken place in the animal tissues, and this process had become more acute after 7 days of treatment. As necrosis is an extrinsically mediated process, release of zooxanthellae was a consequence of damage caused by the hyperthermic stress and not controlled by the host. During this experimentally induced bleaching, individual zooxanthellae undergoing in situ degradation within the same host tissues seemed to be due to both necrosis and PCD. However, there was a lag between the onset of cellular degradation processes of animal tissues and zooxanthellae. Cellular degradation was apparent in the animal tissues, not in zooxanthellae of 4-day-treated anemones but also in zooxanthellae after 7 days of treatment. Whereas host cells suffered necrosis, and this event was the responsible for the release of normal healthy zooxanthellae into the coelenteron, degradation of zooxanthellae resulted from two forms of cell death occurring simultaneously. They were identified as PCD and cell necrosis. Later on, the same authors[37] observed that when exposed to different temperatures ranging from 26 to 34°C the frequency of cell death was greater in the dinoflagellate than in the host, and both types of morphologies, necrotic and apoptotic, increased in frequency throughout the experiment (Fig. 2). Moreover, the frequency of apoptotic morphology of the zooxanthellae was greater than that of necrotic morphology at all stages. It seems that the rapid initial burst of apoptosis-like cell death of host cells in the anemone was not sustained. This gave way to necrotic cell death over time because PCD was initiated as a stress response. Yet, the cells were unable to sustain it during prolonged heat stress due to progressive protein degradation, membrane disruption and energy depletion. The mechanisms controlling the shift between different forms of cell death in response to stress may be highly conserved. However, I wonder whether the morphology of the cells would be "necrotic-like" PCD rather than classical necrosis, because it seems reasonable to think that the shift between apoptosis and necrosis-like PCD would be easier to explain and energetically less expensive for the cell. Nevertheless, this point remains to be answered by further investigations by testing for the presence or absence of caspase-like activities in this species.

There are other examples of dinoflagellates that also undergo PCD under stress conditions. Fluid shear (or turbulence) induced mortality occurs in the red-tide forming dinoflagellate

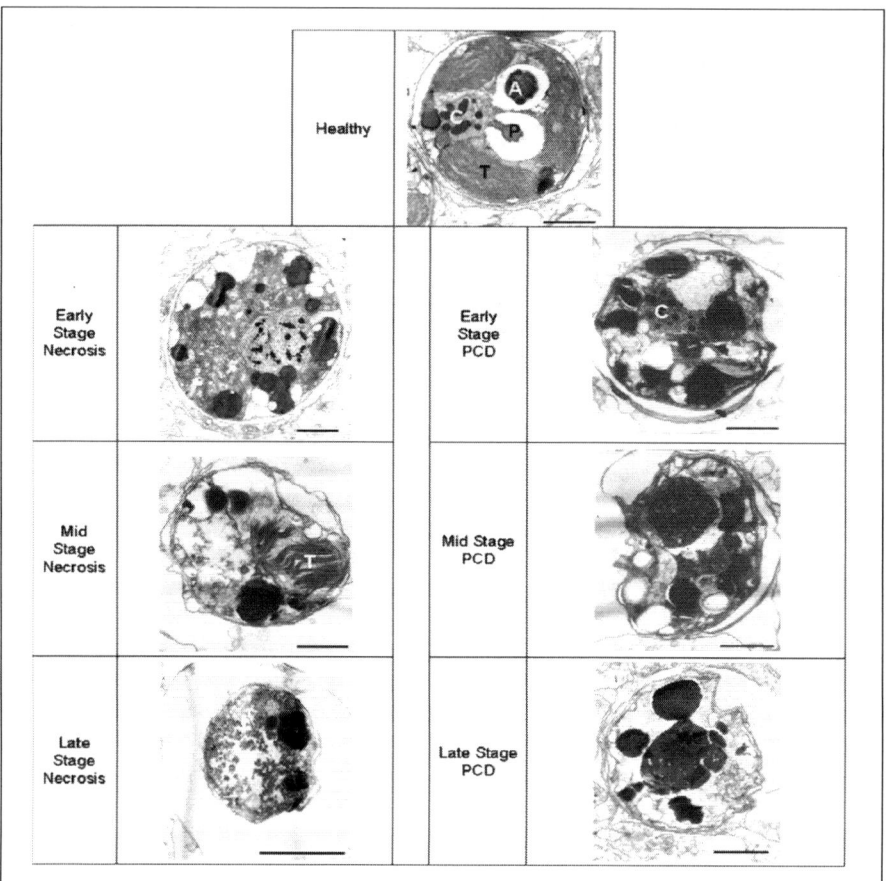

Figure 2. Transmission electron micrographs (TEM) of *Symbiodinium sp.*, the symbiotic zoo-xanthellae of the sea anemone *Aiptasia sp.*, showing the necrotic and apoptotic morphologies. Healthy zooxanthellae are associated with well-defined organelle structure such as thylakoids (T), pyrenoid body (P) and accumulation body (A), as well as chromosomes (C). Within the early stage of necrosis, pyknotic chromatin condensation occurs and chromosomes form irregular clumps throughout the nucleus. Vacuolation of the cytoplasm is evident and organelles appear swollen and distended. Electron-dense aggregate bodies appear in the cytoplasm. At mid-stage necrosis, dilation and rupture of organelles and membranes (e.g., thylakoids; T) occurs and the cytoplasm becomes highly vacuolated. At late stage necrosis cell membranes and walls become ruptured and degraded with little or no evidence of cell contents. The early stages of PCD are characterised by condensation of the cytoplasm and organelles. The nucleus becomes condensed but chromosomes (C) remain intact. As with necrosis, there is an increase in electron-dense aggregate bodies but with little or no vacuolation of the cytoplasm. The periphery of the plasma membrane appears crenated. Mid-stage PCD is characterised by increased crenation of the cell wall and plasma membranes, further condensation of organelles and cytoplasm, larger aggregate bodies and the disappearance of defined chromosomes within the condensed nucleus. At late stage PCD the cell wall becomes 'halo-like' and the aggregate bodies remain the only well-defined cellular components. There is very little or no evidence of membrane rupture during PCD. Scale bars = 2 μm. Reproduced from: SR Dunn et al. Heat stress induces different forms of cell death in sea anemones and their endosymbiotic algae depending on temperature and duration. Cell Death Differ 2004; 11:1-10; ©2004 with permission of The Nature Publishing Group.

Lingulodinium polyedrum. Cells died when shear stress levels were high or when cellular conditions resembled late exponential/stationary phase cultures.[38] Nonswimming cells were dead and presented a distinctive morphology. The theca was intact surrounding a remnant cytoplasm with remaining chlorophyll. DNA fluorescent staining revealed that cells lacked the nucleus, and their occurrence was coincident with an increase in intracellular ROS production. Although authors do not exclude the possibility of a necrotic pathway, they support that the process observed in *L. polyedrum* was consistent with PCD.

Different to the apoptotic, necrotic and necrotic-like PCD morphologies is the paraptotic morphology. A perfect example to show this is the dinoflagellate *Amphidinium carterae*, in which cell death in darkness, and during culture senescence, results in morphological changes without definitive apoptotic characteristics but paraptotic features.[24] When cultures of *A. carterae* where transferred from light to darkness, cell division was arrested and the decline in photosynthetic capacity indicated that cell lysis was a result of dark-induced mortality. Transmission electron microscopy (TEM) revealed cellular dismantling characterized by nonspecific organelle loss, while cells in light had healthy organelles with pyrenoids, starch, condensed chromosomes and intact nuclear and cell membranes. After some days of light deprivation, many cells presented vacuolization and the intracellular content was disorganized, but still, cell membranes appeared intact (Fig. 3). As pointed out previously, some dinoflagellates are thecated (a great difference with metazoan

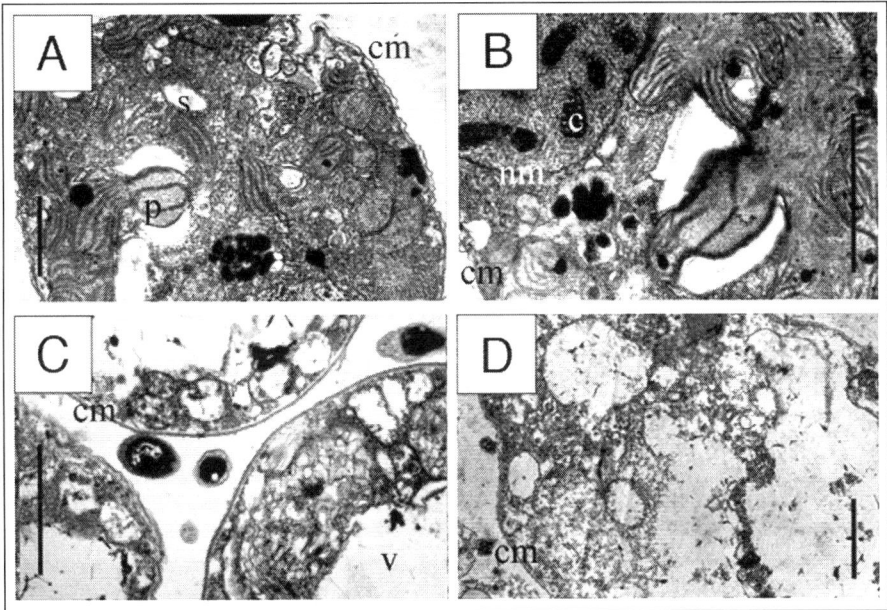

Figure 3. TEM micrographs of *Amphidinium carterae* showing the paraptotic morphology. A) A single *A. carterae* healthy cell in light showing typical vegetative cell morphology including a starch-cap encased pyrenoid (p), thylakoid membranes (t), and starch grains (s). The cell membrane (cm) is also visible in this image, outlining the cel. B) A higher magnification of the image within a healthy cell, the nuclear membrane (nm) encloses regions of darkly-stained condensed chromosomes (c), where again, the cell membrane (cm) is visible. C) Three degenerating cells lying alongside one another after some days in darkness: vacuoles (v) have appeared, although some cell material is still present around the margins of the cell. D) The vast majority of cells contain only amorphous material and vacuoles (v). Scale bar = 1 μm. Reproduced from: Franklin DJ, Berges JA. Mortality in cultures of the dinoflagellate *Amphidinium carterae* during culture senescence and darkness. Proc R Soc Lond B 2004; 271:2099-2107; ©2004 with permission from The Royal Society.

cells) and this feature may account for the lack of cell expansion as the cell dies. However, as time in darkness was prolonged, cells were considerably vacuolated, contained amorphous intracellular material and the chlorotic appearance was obvious during the period of dark-induced mortality. Chromatin condensation, nuclear disintegration and positive TUNEL labelling preceded lysis in some phytoplanktonic species belonging to other different taxonomic groups than dinoflagellates such as *D. tertiolecta*,[31] and *Trichodesmium sp.*[32] Despite DNA fragmentation demonstrated by TUNEL or ISEL end labelling per se is not conclusive to assess PCD, the dinofalgellates *P. gatunense*[22] and *Symbiodinium sp.*[23] showed positive labelling. By contrast, in *A. carterae* TUNEL labelling was negative. It is suggested that the rapid vacuolization, together with the loss of internal structure and the lack of DNA fragmentation support the involvement of a paraptotic process in *A. carterae*.[24] It is remarkable that nucleosomal laddering of DNA has not been detected in any of the species that were TUNEL-positive, possibly because dinoflagellates do not have the typical nucleosomal arrangement observed in other eukaryotes, which is consistent with the absence of ladders in yeast apoptosis.[39] Figures 2 and 3 show the different morphologies of PCD described in dinoflagellates. As we can see, in some cases it is difficult to assign the morphological features displayed to a particular well established cell death programme. Thus, the type of PCD that dinoflagellates undergo could well be species-specific and/or dependent on the triggering factor.

Signalling the Pathway: The Oxidative Burst

The cell death program can be subdivided into three functionally different phases: a stimulus-dependent induction phase, an effector phase during which the wide range of death-stimuli are translated to a central coordinator (the mitochondrion), and a degradation phase during which the alterations commonly considered to define PCD become apparent.[40] Reactive oxygen species (ROS) serve as signal transducers of cell death: firstly, they are the signal during the induction phase, and secondly, they are produced because of mitochondrial permeability transition leading to the final destruction of the cell. Therefore, the oxidative burst is a rapid transient production of ROS such as superoxide (O_2^-), hydrogen peroxide (H_2O_2) or hydroxyl radicals ($OH^•$). The appearance of ROS is known to orchestrate the activation of PCD in animals[41,42] and a variety of responses that involves activation of a plant-encoded pathway for PCD in vascular plants.[43-45] ROS are also known to mediate PCD in unicellular organisms such as kinetoplastids,[46,47] yeast[39] and chlorophytes (Segovia and Berges, unpublished data). In marine environments, oxidative stress in algae is often caused by high light intensity and increased pH of the seawater as a result of algal photosynthesis, as well as by nutrient and carbon limitation.[16] To counteract the production of ROS cells are provided with antioxidant enzymatic activities and compounds such as superoxide dismutase (SOD), catalase, peroxidases, ascorbic acid, glutathione, mycosporine-like amino acids and carotenoids among others.[48] Particularly, SOD exists in eukaryotic protists, and specifically in dinoflagellates.[49-51] With regard to ROS, it is not always clear that their appearance is linked to PCD and there are a number of reports that describe its accumulation in several species under different conditions. For instance, *A. carterae* when exposed to toxic levels of copper, accumulated ROS and produced metabolic changes leading to cell death.[52] *L. polyedrum* also generated ROS under high irradiances, provoking lipid peroxidation and cell death.[53] In the same species, chlorinated phenolics or heavy metals induced the formation of ROS and an increase in ascorbate peroxidase and SOD that ended up in cell death.[54] Fluid shear also lead to ROS formation and cell death in *L. polyedrum*.[38] In these examples cell death was observed but not a clear type was assessed.

Then, dinoflagellates produce ROS under different conditions but the link between ROS accumulation and PCD is not clear yet and disserves further investigation. However, there are a few good examples to illustrate what is known up to date about the relationship between ROS and PCD in dinoflagellates. Some of the cases are zooxanthellae on which I will focus later on in this chapter (see "coral bleaching"). Other is the bloom-forming freshwater species *P. gatunense* from Lake Kinneret (Israel).This dinoflagellate is exposed to different environmental factors during spring, so diel and seasonal responses to environmental stimuli are common. An antioxidant mechanism was found to be operative during the spring bloom.[51] The bloom terminated suddenly in summer when the pH raised and the availability for dissolved CO_2 was drastically reduced. Antioxidant enzymatic activities and compounds such as ascorbate and catalase, which are in charge of eliminating the H_2O_2 produced photosynthetically, were present at high rates. Such activities accumulated at the end of the bloom, and seemed to show a diel cycle coincident with peak mid-day irradiances. Later on, the explanation for this came up.[22] The CO_2 limitation that develops during the bloom, diverts photosynthetic electrons from CO_2 fixation to oxygen, resulting in the production of ROS at a concentration enough to trigger PCD. To check whether ROS were the mediators for the decline of the bloom in summer, they stained both cells from the lake and from culture with a fluorescent dye that passively diffuses across most cell membranes where it is oxidized localizing in the mitochondria. The percentage of ROS-positive *Peridinium* cells increased during growth in the lake (Fig. 4A) and in batch cultures. Simultaneously, the concentration of H_2O_2 in the cultures also ascended and was accompanied by a rising of antioxidant activities. In parallel, DNA fluorescence, measured by using Sytox -a membrane impermeable DNA binding fluorescent dye- increased considerably (Fig. 4B). DNA suffered fragmentation as demonstrated by TUNEL labelling and the central role of ROS as a trigger of PCD in *P. gatunense* was demonstrated.

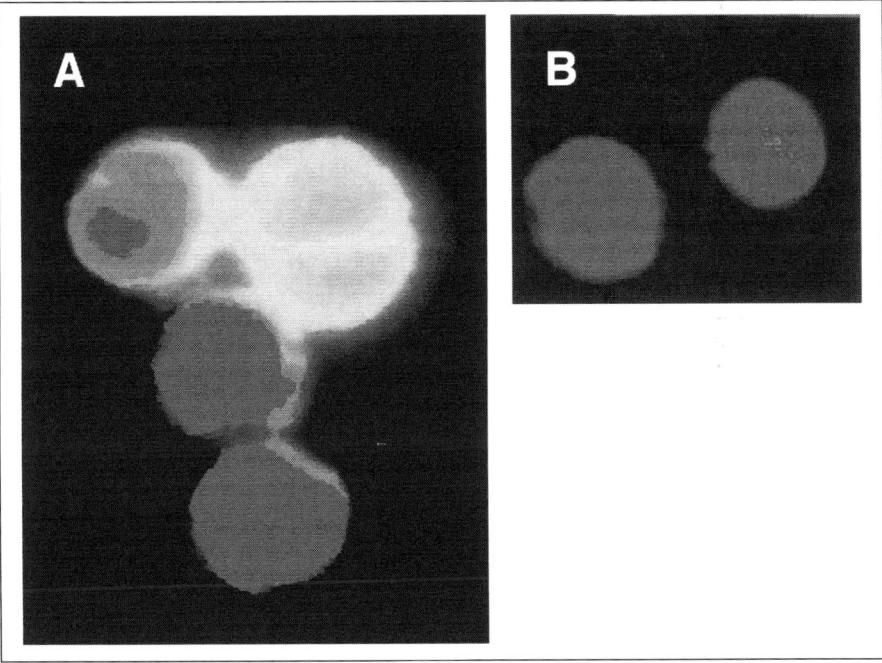

Figure 4. Cell death and accumulation of ROS in *P. gatunense*. ROS were detected by using dihydrorhodamine 123. This dye passively diffuses across most cell membranes where it is oxidized to cationic rhodamine 123, which localizes in the mitochondria. The dye does not directly detect superoxide, but rather reacts with hydrogen peroxide in the presence of peroxidase, cytochrome c or Fe^{2+}. A) The darker color is due to the autofluorescence of the chlorophyll, whereas clearer color indicates ROS-containing cells. Cells emitting green fluorescence were swimming, albeit slower than the unstressed ones. B) Cell death was assayed with Sytox, which is a high affinity nucleic acid stain that easily penetrates cells with compromised plasma membranes and yet will not cross the membranes of live cells. Sytox may therefore indicate an early stage in cell death. The image shows a stained clear nucleus in he right-hand cell. Reprinted from: Assaf Vardi et al. Programmed cell death of the dino-flagellate *Peridinium gatunense* is mediated by CO_2 limitation and oxidative stress. Curr Biol 1999; 9:1062; ©1999 with permission from Elsevier.

Yet, ROS are not the only signals that intervene during PCD. Many different signal transduction pathways participate in the induction phase of PCD. Beyond this stage, mitochondrial permeability transition is the central coordinator of PCD deciding whether a cell will die.[55] Therefore, during the effector phase, universal regulatory events coming from the mitochondrion, including the Bcl-2 family members, take place. The Bcl-2 family members are mitochondrial proteins. Some, such as Bcl-2, Bcl-XL, and BAG-1, inhibit PCD, while others, such as Bax and Bak, promote PCD.[56,57] They serve as critical regulators of signalling pathways involved in apoptosis because they can interact with each other through a complex network of homo- and heterodimers, with one monomer antagonizing or enhancing the function of another. In this way, the ratio of pro- and antiapoptotic Bcl-2 family proteins in a cell may determine the likelihood of the cell undergoing PCD.

The problem is that currently, no such genes have been identified in plants or unicellular eukaryotes in spite of the genome databases that are available. Nonetheless, some recent evidence argues for the existence of an evolutionarily conserved pathway for the control and execution of PCD in animal and plant cells.[58] Even in silico studies of Bcl-2 family proteins and other regulators, carried out in *Plasmodium falciparum*, the best possible candidate among alveolates to find something, failed to detect potential orthologues of these mammalian proteins.[59] We are looking forward for further research to unravel the existence and participation of these components in protists' PCD.

Ecological Relevance of ROS: Coral Bleaching

One of the most relevant consequences of ROS accumulation is coral bleaching -although we shall see that bleaching is not restricted to corals-. Thermal stress is the main cause for bleaching through the generation of ROS around the photosystem II (PS II), diminishing the efficiency of the photosynthetic apparatus.[60,61] The oxidative stress has been proposed as a unifying mechanism for several environmental injuries that cause this phenomenon. Bleaching, explicitly, is a stress response in which any animal host cell -mainly from the phylum Cnidaria, i.e., sea anemones, corals and jellyfish- in symbiosis with dinoflagellates from the genus *Symbiodinium*, lose their zooxanthellae either by exocytosis, pinching off, host cell detachment, necrosis or apoptosis.[62] Therefore, PCD is one of the mechanisms by which zooxanthellae are released outwards these organisms. Simultaneously to the reduction of zooxanthellae, algal pigments are lost as well. This episode occurs when the average summer temperature in the seawater suffers an increase of 2-3°C as consequence of global climate change, principally due to the emission of greenhouse gases such as CO_2 and CH_4.[63-65]

Heat stress is the principal cause of coral bleaching, but there are other factors such as high solar irradiance or anthropogenic, that also trigger bleaching. The extent of bleaching is directly related to the magnitude of temperature elevation and the duration of exposure. Interestingly, it appears that there is a critical threshold temperature separating thermally tolerant from sensitive species of zooxanthellae, and this is determined by the lipid composition of the thylakoidal membranes.[66,67] Several mechanisms have been proposed to explain the pathways by which bleaching goes through, but it is not clearly known whether they occur simultaneously, sequentially or if they are mutually exclusive. Among these events, the D1 protein from the PS II is damaged by ROS[68] and the Calvin cycle is disrupted, hence CO_2 fixation is arrested.[69] There are changes in the pattern of protein phosphorylation leading to animal cell detachment and loss.[70] Accompanying these events, PCD or necrosis is established and cell lysis overcomes.[23] Why zooxanthellae are expulsed or dead-and-released at a particular moment? All symbiotic organisms are very complex and display relationships that are in the borderline between parasitic and mutualistic. In the nonbleaching symbiosis, there is a harmonious maintenance between the cell growth rates of both partners, i.e., the host (cnidarians) and the endosymbiont (zooxanthellae).[71] However, this situation varies when a stressor appears on screen. Under thermal stress, for instance, the algal cells are metabolically compromised with a reduced capacity to provide the animal with photosynthetic carbon.[72] This raises the possibility that the symbiosis is maintained by the sustained production by the algal cells of a signal that inactivates the animal defence responses to foreign organisms, and the loss of this signal drives to bleaching. A candidate for this putative signal is the reception of algal-derived photosynthates by the animal. *Symbiodinium* cells metabolically unbalanced by a trigger of bleaching may release specific compounds or display relatively uncontrolled leakage of cytoplasmic contents, to which the animal partners responds defensively.[73] Accordingly, trigger mechanisms may come from the host via intercellular signalling and PCD of zooxanthellae may also be initiated by mechanisms involved in potential host digestion or could be mediated by the algal cell itself. In this sense, the capacity of algal cells to mediate their own cell death may allow the host to survive episodes of bleaching.[74]

Recently, a number of reports have tried to clarify the underlying mechanism for the sequence of events induced by bleaching and the role of ROS at the molecular level. Franklin et al[74] tested whether dinoflagellate cell death preceded bleaching in the coral *Stylophora pistillata*. Zooxanthellae death preceded bleaching of the coral, and was coincident with both depressed photosynthesis, and increased intracellular ROS accumulation. Concurrently, TEM revealed the accumulation body material in dinoflagellate cells. The accumulation body is thought to be an autophagic structure responsible for the breakdown of membranous structures and organelles into their basic components,[75] and represents the induction of autolytic cell death.[76,77] Thus, the accumulation body could well be functionally equivalent to the apoptotic bodies formed during the PCD of metazoan cells.[23] For this reason, it appears that ROS could act as a cell death trigger and symbiont cell death is likely to be one of the first steps of coral bleaching. The coral *Montastraea faveolata*[78] exposed to the synergistic effect of solar radiation on thermal stress provides an essential biochemical and molecular insight in the process. PS II fluorescence was depressed and the concentration of the D1 protein was very low. Antioxidant enzymatic activities such as SOD were present, indicating that the production of ROS had increased[79] and was the responsible for the mentioned effects. DNA damage was checked by detecting cyclobutane pyrimidine dimmers by western blotting. DNA damage is caused indirectly by exposure to ultraviolet radiation (present in the solar radiation) through the photodynamic production of ROS, and can lead to PCD if not repaired.[80] One of the key cell cycle genes activated after DNA damage is p53 that initiates the downstream expression of cell cycle genes that cause cells to arrest in G1/S while DNA damage is being repaired.[81] If DNA repair is not possible, then p 53-mediated apoptosis may be initiated. The p53 protein was detected in *M. faveolata* by immunoblotting, and the expression pattern of the putative p53 was consistent with the pattern of DNA damage. Based on the ultrastructural morphological evidence that apoptosis and necrosis occur in thermally stressed symbiotic cnidarians, as we have already seen,[23] and that a putative p53 protein is up-regulated in response to DNA damage, the data we have at present significantly supports the occurrence of apoptosis and possibly cell necrosis mediated by ROS in thermally stressed symbiotic cnidarians. ROS are likely to be involved at both ends of the sequence of events comprising from photosynthetic and DNA damage to apoptosis and cell necrosis.

From host controlling zooxanthellae population growth[82] to the adaptive bleaching hypothesis,[83] many theories have tried to explain why the *Symbiodinium* symbioses persists although bleaching is apparently deleterious. Whatever the answer is, the variability in the bleaching response that comes about under different environmental conditions is extremely complicated due to the genetic diversity of algal symbionts within and between different cnidarians.[66,67,84,85] This has important connotations and repercussions because of ecological collapse of the world's reefs, and subsequently loss of biodiversity and entire ecosystems, caused largely by global warming and the ozone layer depletion. The lack of a proper reef management strategy results in coral reefs, most likely the most productive ecosystem on earth, disappearance.

Metacaspases: Killers of the Cell

We have seen in the previous sections that signalling mechanisms mediate the effector phase of PCD. Following this, is the degradation period during which the alterations commonly considered to define PCD become apparent. The celebrities of this phase are the proteases. Proteolysis is a common mechanism that regulates a number of events within the cell such as growth, gene expression, development and PCD among others.[86] This fact makes of proteolysis the appropriate process for cellular adaptation to changing environments. Proteases in general are kept quiescent in the cell, lurking and poised to commit the cell to a particular fate. In the case of PCD, the proteases in charge of executing the cell are the caspases (although there are caspase-indepent forms of PCD).[34,87] These proteins exist originally as a zymogen or proenzyme. The N-terminus exerts regulatory and activating functions. The catalytic region is constituted by two domains of 20 and 10 kD, which upon activation will yield the two subunits of the active enzyme. Depending on the length of the N-terminus caspases are divided in two groups: caspases 1, 2, 4, 5, 8, 9 and 10 have a long prodomain acting upstream the cascade, regulating and activating other caspases. Caspases 3, 6 and 7 have a short prodomain and act downstream the cascade after being activated by upstream caspases. Caspases are among the most specific proteases, having an unusual and stringent requirement for cleavage after aspartic acid. Recognition of at least four canonical amino acids, NH_2-terminal to the cleavage site, is also a requirement for efficient catalysis; very few proteolytic enzymes can cleave the fluorogenic substrates used to detect caspases.[88] The caspase family includes not only caspases but also other members called paracaspases and metacaspases. Metacaspases were discovered in yeast, plants, protists and some bacteria by iterative PSI-BLAST searches.[89] Paracaspases were identified through their remote homology with caspases and metacaspases, and are found in animals and slime moulds. The homology of metacaspases to human caspases is not restricted to sequences surrounding the catalytic residues but also extends into the secondary structure. Nine metacaspases have been identified in the vascular plant *Arabidopsis* genome, two of which are upregulated by bacterial pathogens.[58]

The story changes a little bit when we talk about phytoplankton in general and about dinoflagellates in particular. The data about metacaspases in protists are scarce. Caspase-like enzymatic activities and caspase immunodetection were reported for the first time in the unicellular cholorophyte *D. tertiolecta*[31] when the authors observed that the fluorogenic caspase-specific substrates WEHD (caspase 1), DEVD (caspase 3), IETD (caspase 8) and LEHD (caspase 9) were cleaved and that the activities were inhibited by specific caspase inhibitors. Currently there are very few reports about them in dinoflagellates. One of the reports mentions the enzymatic activity of a possible cysteine protease involved in PCD.[22] The scenario is the same as the one described above to explain the link between ROS and PCD in *P. gatunense*. We had just seen that CO_2 limitation resulted in the generation of ROS to an extent that induced PCD in this species. Nevertheless, PCD was avoided when cells were treated with E-64, an inhibitor of cysteine proteases. The inhibitor completely suppressed ROS-and Sytox-positive cells, and inhibited cell death following treatment with H_2O_2. Interestingly, the inhibitor stimulated cyst formation in cells collected from the lake. The causes that determine the fate of *Peridinium* cells, i.e., cyst formation or death, are poorly understood. The authors suggest that E-64 probably intervened in the PCD pathway inducing the cells to form cysts. This is coincident with the implication of cysteine proteases genes and cell differentiation in the ciliated protist *Sterkiella histriomucorum*. In this case, gene expression is involved with excystment instead of cystment.[90] Neither specific caspase substrates nor inhibitors were tested in *Peridinium*, nor were proteins western blotted by using caspase antibodies. Nonetheless, mitogen-activated protein kinases (MAPK) seemed to be involved in the regulation of the oxidative burst.[91] Okamoto and Hastings[92] found important redox-regulated genes, when they were identifying genes involved in the response of the dinoflagellate *Pyrocystis lunula* to ROS and reactive nitrogen species (RNS)-generating agents, such as paraquat (an herbicide) and $NaNO_2$, by using microarrays. *P. lunula* cells show a very typical half-moon shape. When the cells were exposed to the oxidative burst-generating compounds, they changed their shape and became spherical. The authors identified members of the ubiquitin system, which revealed increased transcript levels for E2 ubiquitin-conjugating enzyme, E3 ubiquitin-protein ligase, and subunit p28 of the 26s proteasome in $NaNO_2$-treated cells. But, what was most important, they found one gene encoding a metacaspase that was upregulated by nitrite. Indeed, RNS are able to induce apoptosis through activation of p53 and caspases in megakaryocytes[93] and nitric oxide (NO) is known to mediate PCD in plants.[94,95] However, after treatment with the RNS-generating agent, most or all spherical cells recovered their normal shape and started to grow again. One effect of metazoan caspases is the activation of certain target proteases that cleave actin, affecting the cytoskeleton. Metacaspases may act similarly, and therefore spherical cells were formed again. If metacaspases have caspase activity and regulate PCD in unicellular protists, this would indicate that metacaspases represent an ancestral core of PCD proteins.[33] However, although their activity has been measured in some unicellular phytoplankters,[31,32] the genes have only been characterised in silico and the enzymatic activity associated to these genes has not been demonstrated yet. Analyses of completed genome sequences of prokaryotic and eukaryotic phytoplankton have revealed the widespread presence of metacaspases in some cyanobacteria, in the unicellular eukaryotic chlorophyte *Chlamydomonas reinhardtii*, in the marine diatom *Thalassiosira pseudonana* and in the marine haptophyte *Emiliania huxleyi*. Metacaspases from these eukaryotic phytoplankton lineages cluster within a group of metacaspases that includes unicellular protists, such as fungi, trypanosomes and vascular plants, which might indicate that they have similar physiological roles.[33] This supports the idea that the dinoflagellates caspases are metacaspases, as the one found in *P. lunula*.

Where Did the PCD Genes Come From?

With regard to PCD evolution, the chapter by Pérez et al in this book, is fully devoted to it (see "PCD in protozoa: an evolutionary point of view") so I will just mention a few issues about dinoflagellates with this respect, and always within the frame of phytoplankton evolution.

Apparently, the cell death process in unicells confers no obvious ecological or evolutionary advantage to the organism, and cell death can be seen as a maladaptive strategy for unicells, although there are few different theories to explain this when referred to phytoplankton.[33] It has been hypothesized that key elements of cell death pathways were transferred to the nuclear genome of early eukaryotes through ancient viral infections in the Precambrian Ocean,[27] prior to the evolution of multicellular organisms, and were subsequently appropriated in both metazoan and higher plant lineages.[31,96] However, I have tried to identify potential orthologs of the PCD components mostly in the phylum Dinophyta and in the alveolates, using standard protein-protein BLAST (blastp), PSI-BLAST, searching the conserved domain and the protein homology by domain databases. The apoptosis-inducing factor (AIF) and endonuclease G threw positive identity results in cyanobacteria and apicomplexans. Bcl-2 did it to putative proteins of *P. lunula*, *L. polyedrum* and *Heterocapsa triquetra*. The apoptotic protease-activating factor 1 (Apaf-1) showed identities to proteins of *Heterocapsa pygmaea* and *H. triquetra*, ciliates, apicomplexans, kinetoplastids (euglenozoa) and cyanobacteria. Unfortunately, although the searches threw positive identity results, they were non significant. Only metacaspases were significant (see the metacaspases section, further up in this chapter).

Returning to the main issue of how PCD machinery became present in unicells, the idea of a real lateral transfer of some of the cell death elements occurring from viruses to aerobic bacteria and / or photosynthetic bacteria at some point, which carried them out into the Precambrian eukaryotes, should not be discarded. Some findings argue in favour for this: HaRNAV, a single-stranded RNA virus, is capable to infect the toxic bloom-forming alga *Heterosigma akashiwo*.[97] Other single-stranded RNA viruses, HcRNAVs are able to infect the bivalve-killing dinoflagellate *Heterocapsa circularisquama*.[98] The most compelling evidence though, is the finding of a *Coccolithovirus* (a recently discovered group of viruses that infect the globally important marine calcifying bloom-forming microalga *Emiliania huxleyi*) encoding genes involved in the biosynthesis of ceramide, a sphingolipid known to induce apoptosis, and eight proteases.[99] Thus, we might consider that the transfer of some of the cell death genes probably did not only occur between a virus and the nuclear genome but also occurred when the endosymbiotic event took place 1.5 b.a. between a virus and ancient bacteria before the origin of eukaryotic life itself. If this is true, then endosymbiosis and cell death may have developed in tandem. The endosymbiotic theory[100] postulates that eukaryotic life arose when a prokaryotic host cell engulfed an aerobic bacterium, which ultimately became the mitochondrion. Kroemer[101] pointed out that apoptosis itself may have evolved together with the endosymbiotic incorporation of aerobic bacteria (the precursors of mitochondria) into ancestral unicellular eukaryotes. The apoptotic machinery in eukaryotic phytoplankton lineages might also have been established by an endosymbiotic event with a cyanobacterium, retained in independently evolving superfamilies and subsequently transferred to the nuclear genome of the host.[33] In opposition to this, is the idea supporting that the endosymbiotic bacterial ancestors of mitochondria are unlikely to have contributed to the recent mitochondrial death machinery and therefore, these components may derive from mutated eukaryotic precursors and might have invaded the respective mitochondrial compartments. Although there is no direct evidence, it seems that the prokaryotic-eukaryotic symbiosis created the space necessary for sophisticated death mechanisms on command, which at the end is the prerequisite for multicellularity.[102] Evolution never invents new genes but plays variations on old themes by DNA mutations thus; other possibility is to assume that PCD genes were always there. Ameisen[103] postulates the "original sin" hypothesis, according to which, the origin of the capacity to self-destruct may be as ancient as the origin of the very first cell. Thus, if effectors of the cell survival machinery can also be effectors of the self-destruction of the cell in which they operate, then the requirement for coupling cell survival to the prevention of self-destruction may be as old as the origin of the first cell. In Ameisen's words: "cell suicide is an unavoidable consequence of self-organization".

Nevertheless, arguments in favour of the inheritance of the PCD components through endosymbiotic mechanisms seem plausible (see ref. 33). Phytoplankton evolution comprises both the bacterial and eukaryotic domains of life and at least eight major eukaryotic divisions or phyla.[104] The photosynthetic origin of eukaryotic phytoplankton goes back to the endosymbiotic events that took place between an ancestral mitochondrion-containing eukaryotic cell and cyanobacteria.[100,105,106] Photosynthetic eukaryotes inhabited coastal waters at the beginning of evolution, but reasonably early in time, there was a division leading to two ancient superfamilies, the 'green' and 'red' lineages.[107] Each lineage evolved independently after the primary endosymbiotic event.[100] The green lineage includes the chlorophytes and dominated the eukaryotic marine phytoplankton communities during the Mesoproterozoic and Paleozoic becaming less abundant after the Permian extinction.[104] The red lineage includes diatoms, coccolithophorids and the dinoflagellates, and has come to dominate the modern ocean.[104] Increasingly, studies of genes and genomes are indicating that considerable horizontal transfer has occurred between prokaryotes.[108] Then, it is possible that metacaspase genes were acquired by cyanobacterial genomes through lateral gene transfer, but they could also have been present in the common ancestor of the α-Proteobacteria and cyanobacteria, having undergone some lineage-specific differential loss.[33] The cyanobacteria have considerable diversity in the phylogenetic analysis of the caspase-like protein family, forming several distinct clusters thus, it is assumed that lateral gene transfer has been a key element in their evolution.[33] The conclusion is that the presence of eukaryotic cell-death domains in the genomes of Proteobacteria, indicates that these genes have a bacterial origin. The transfer of such genes occurred during the endosymbiotic events but possibly they are viral in origin.

Conclusions

PCD in unicells is difficult to explain because is a mechanism which offers negative selective pressure, therefore must be maladaptative in origin. In spite of this, cells have managed to use PCD for several ecologically relevant purposes. Some of these drivers are: growth control of the cell population as it happens in corals; population benefit according to which damaged cells are eliminated; arresting viral infections by host cell death; allelopatic interactions and niche specialization to favour the growth of a particular species; and genetic fitness or altruism by which some cells die in benefit of the whole population. PCD in phytoplankton is truly the result of a programme activated by environmental factors, then it is important to understand how activation happens and what occurs within the cell in response. The existence of genetically driven cell death phenomena in phytoplankton has important implications for species successions and biogeochemical cycling in aquatic ecosystems, and finally leads in the biodiversity of the oceans and freshwaters systems.

Acknowledgements

I thank Dr. Maria Faust at the National Museum of Natural History, Smithsonian Institution, Washington DC, USA, for providing the SEM micrograph of *L. polyedrum*. I thank Dr. Carlos Jimenez for comments on the MS.

References

1. Graham LE, Wilcox LW. Algae. Upper Saddle River: Prentice-Hall, 2000.
2. Taylor FJR. The biology of dinoflagellates. In: Taylor FJR, ed. General Group Characteristics. Oxford: Blackwell, 1985:1-24.
3. Bhattacharya D, Yoon HS, Hackett JD. Photosynthetic eukaryotes unite: Endosymbiosis connects the dots. Bioessays 2004; 26:50-60.
4. Murray S, Flø Jørgensen M, Hoc SYW et al. Improving the analysis of dinoflagellate phylogeny based on rDNA. Protist 2005; 156:269-286.
5. Livolant F, Bouligand Y. New observations on the twisted arrangement of dinoflagellate chromosomes. Chromosoma 1978; 8:21-44.
6. Bendich AJ, Drlica K. Prokaryotic and eukaryotic chromosomes: What's the difference? Bioessays 2000; 2:481-486.
7. Sala-Rovira M, Geraud ML, Caput D et al. Molecular cloning and immunolocalization of two variants of the major basic nuclear protein (HCc) from the histone-less eukaryote Crypthecodinium cohnii (Pyrrhophyta). Chromosoma 1991; 100:510-518.
8. Wong JT, New DC, Wong JC et al. Histone-like proteins of the dinoflagellate Crypthecodinium cohnii have homologies to bacterial DNA-binding proteins. Eukaryot Cell 2003; 2:646-650.
9. Fagan TF, Li JF, Chudnovsky J et al. Cloning, sequencing and expression of a histone-like protein from the photosynthetic dinoflagellate Gonyaulax polyedra. J Phycol 2000; 36:21-22.
10. Chudnovsky Y, Li JF, Rizzo PJ et al. Cloning, expression, and characterization of a histone-like protein from the marine dinoflagellate Lingulodinium polyedrum (Dinophyceae). J Phycol 2002; 38:543-550.
11. Hackett JD, Scheetz TE, Yoon HS et al. Insights into a dinoflagellate genome through expressed sequence tag analysis. BMC Genomics 2005; 6:80.
12. Dodge JD. Dinoflagellate with both a mesocaryotic and a eucaryotic nucleus: Fine structure of nuclei. Protoplasma 1971; 73:145-157.
13. Costas E, Goyanes V. Ultrastructure and division behavior of dinoflagellate chromosomes. Chromosoma 1987; 95:435-441.
14. Wong JTY, Kwok ACM. Proliferation of dinoflagellates: Blooming or bleaching. Bioessays 2005; 27:730-740.
15. Guillebault D, Sasorith S, Derelle E et al. A new class of transcription initiation factors, intermediate between TATA box-binding proteins (TBPs) and TBP-like factors (TLFs), is present in the marine unicellular organism, the dinoflagellate Crypthecodinium cohnii. J Biol Chem 2002; 277:40881-40886.
16. Falkowski PG, Raven JA. Aquatic Photosynthesis. Malden, Massachusetts: Blackwell Science, 1997.
17. Anderson DM. Toxic red tides and harmful algal blooms: A practical challenge in coastal oceanography. Rev Geophys 1995; 33:1189-1200.
18. Richardson K. Harmful or exceptional phytoplankton blooms in the marine ecosystem. Adv Mar Biol 1997; 31:302-385.
19. Sellner KG, Doucette GJ, Kirkpatrick GJ. Harmful algal blooms: Causes, impacts and detection. J Microbiol Biotechnol 2003; 30:383-406.
20. Walsh JJ. Death in the sea: Enigmatic phytoplankton losses. Prog Oceanogr 1983; 12:1-86.
21. Fogg GE, Thake B. Algal cultures and phytoplankton ecology. University of Wisconsin Press, 1987.
22. Vardi A, Berman-Frank I, Rozenberg T et al. Programmed cell death of the dinoflagellate Peridinium gatunense is mediated by CO_2 limitation and oxidative stress. Curr Biol 1999; 9:1061-1064.
23. Dunn SR, Bythell JC, Le Tissier MDA et al. Programmed cell death and cell necrosis activity during hyperthermic stress-induced bleaching of the symbiotic sea anemone Aiptasia sp. J Exp Mar Biol Ecol 2002; 272:29-53.
24. Franklin DJ, Berges JA. Mortality in cultures of the dinoflagellate Amphidinium carterae during culture senescence and darkness. Proc R Soc Lond B 2004; 271:2099-2107.
25. Cornillon S, Foa C, Davoust J et al. Programmed cell-death in Dictyostelium. J Cell Sci 1994; 107:2691-2704.
26. Ameisen JC. The origin of programmed cell death. Science 1996; 272:1278-1279.
27. Berges JA, Falkowski PG. Physiological stress and cell death in marine phytoplankton: Induction of proteases in response to nitrogen or light limitation. Limnol Oceanogr 1998; 43:129-135.
28. Frohlich KU, Madeo F. Apoptosis in yeast - A monocellular organism exhibits altruistic behaviour. FEBS Lett 2000; 473:6-9.

29. Lewis K. Programmed cell death in bacteria. Microbiol Mol Biol Rev 2000; 64:503-514.
30. Ning SB, Guo HL, Wang L et al. Salt stress induces programmed cell death in prokaryotic organism Anabaena. J Appl Microbiol 2002; 93:15-28.
31. Segovia M, Haramaty L, Berges JA et al. Cell death in the unicellular chlorophyte Dunaliella tertiolecta: An hypothesis on the evolution of apoptosis in higher plants and metazoans. Plant Physiol 2003; 132:99-105.
32. Berman-Frank I, Bidle KD, Haramaty L et al. The demise of the marine cyanobacterium, Trichodesmium spp., via an autocatalyzed cell death pathway. Limnol Oceanogr 2004; 49:997-1005.
33. Bidle KD, Falkowski PG. Cell death in planktonic, photosynthetic microorganisms. Nature Rev Microbiol 2004; 2:643-655.
34. Kitanaka C, Kuchino Y. Caspase-independent programmed cell death with necrotic morphology. Cell Death Differ 1999; 6:508-515.
35. Sperandio S, de Belle I, Bredesen DE. An alternative, nonapoptotic form of programmed cell death. Proc Natl Acad Sci USA 2000; 97:14376-14381.
36. Wyllie AH, Golstein P. More than one way to go. Proc Natl Acad Sci USA 2001; 98:11-13.
37. Dunn SR, Thomason JC, Thissler ML et al. Heat stress induces different forms of cell death in sea anemones and their endosymbiotic algae depending on temperature and duration. Cell Death Differ 2004; 11:1213-1222.
38. Juhl AR, Latz MI. Mechanisms of fluid shear-induced inhibition of population growth in a red-tide dinoflagellate. J Phycol 2002; 38:683-694.
39. Madeo F, Frohlich E, Ligr M et al. Oxygen stress: A regulator of apoptosis in yeast. J Cell Biol 1999; 145:757-767.
40. Jabs T. Reactive oxygen intermediates as mediators of programmed cell death in plants and animals. Biochem Pharmacol 1999; 57:231-245.
41. Cohen GM. Caspases: The executioners of apoptosis. Biochemical J 1997; 326:1-16.
42. Leist M, Nicotera P. The shape of cell death. Biochem Biophys Res Com 1997; 236:1-9.
43. Pennell RI, Lamb C. Programmed plant cell in plants. Plant Cell 1997; 9:1157-1168.
44. Lam E, Kato N, Lawton M. Programmed cell death, mitochondria and the plant hypersensitive response. Nature 2001; 411:848-853.
45. Chichkova NV, Kim SH, Titova ES et al. A plant caspase-like protease activated during the hypersensitive response. Plant Cell 2004; 16:157-171.
46. Ridgley EL, Xiong ZH, Ruben L. Reactive oxygen species activate a Ca^{2+}-dependent cell death pathway in the unicellular organism Trypanosoma brucei. Biochem J 1999; 340:33-40.
47. Sen N, Das BB, Ganguly A et al. Camptothecin-induced imbalance in intracellular cation homeostasis regulates programmed cell death in unicellular hemoflagellate Leishmania donovani. J Biol Chem 2004; 279:52366-5237.
48. Lesser MP. Oxidative stress in marine environments. Biochem Physiol Ecol Annu Rev Physiol 2006; 68:253-78.
49. Lesser MP, Shick JM. Effects of irradiance and ultraviolet radiation on photoadaptation in the zooxanthellae of Aiptasia pallida: Primary production, photoinhibition, and enzymatic defences against oxygen toxicity. Mar Biol 1989; 102:243-55.
50. Hollnagel HC, di Mascio P, Asano CS et al. The effect of light on the biosynthesis of β-carotene and superoxide dismutase activity in the photosynthetic alga Gonyaulax polyedra. Brazil J Med Biol Res 1996; 29:105-10.
51. Butow BJ, Wynne D, Tel-Or E. Superoxide dismutase activity in Peridinium gatunense in Lake Kinnert: Effect of light regime and carbon dioxide concentration. J Phycol 1997; 33:787-93.
52. Lage OM, Sansonetty F, O'Connor JE et al. Flow cytometric analysis of chronic and acute toxicity of copper (II) on the marine dinoflagellate Amphidinium carterae. Cytometry 2001; 44:226-235.
53. Cardozo KHM, de Oliveira MAL, Tavares MFM et al. Daily oscillation of fatty acids and malondialdehyde in the dinoflagellate Lingulodinium polyedrum. Biol Rhythm Res 2002; 33:371-381.
54. Leitao MAD, Cardozo KHM, Pinto E et al. PCB-induced oxidative stress in the unicellular marine dinoflagellate Lingulodinium polyedrum. Arch Environ Con Tox 2003; 45:59-65.
55. Kroemer G, Petit P, Zamzami N et al. The biochemistry of programmed cell-death FASEB J 1995; 9:1277-1287.
56. Sato T, Hanada M, Bodrug S et al. Interactions among members of the Bcl-2 protein family analyzed with a yeast two-hybrid system. Proc Natl Acad Sci USA 1994; 91:9238-9242.
57. Korsmeyer SJ. Regulators of cell death. Trends Genet 1995; 11:101-105.
58. Watanabe N, Lam E. Recent advance in the study of caspase-like proteases and Bax inhibitor-1 in plants: Their possible roles as regulator of programmed cell death. Mol Plant Pathol 2004; 5:65-70.
59. Deponte M, Becker K. Plasmodium falciparum - do killers commit suicide? Trends Parasitol 2004; 20:165-169.

60. Nishiyama Y, Yamamoto H, Allakhverdiev SI et al. Oxidative stress inhibits the repair of photodamage to the photosynthetic machinery. EMBO J 2001; 20:5587-5594.
61. Takahashi S, Nakamura T, Sakamizu M et al. Repair machinery of symbiotic photosynthesis as the primary target of heat stress for reef building corals. Plant Cell Physiol 2004; 45:251-255.
62. Gates RD, Baghdasarian G, Muscatine L. Temperature stress causes host cell detachment in symbiotic cnidarians: Implications for coral bleaching. Biol Bull 1992; 182:324-32.
63. Ramanathan V, Collins W. A thermostat in the tropics. Nature 1993; 361:410-411.
64. Hoegh-Guldberg O. Climate change, coral bleaching and the future of the world's coral reefs. Mar Freshw Res 1999; 50:839-66.
65. Lesser MP. Experimental coral reef biology. J Exp Mar Biol Ecol 2004; 300:217- 52.
66. Rowan R. Coral bleaching - Thermal adaptation in reef coral symbionts Nature 2004; 430:742-742.
67. Tchernov D, Gorbunov MY, de Vargas C et al. Membrane lipids of symbiotic algae are diagnostic of sensitivity to thermal bleaching in corals. Proc Natl Acad Sci USA 2004; 101:13531-13535.
68. Warner ME, Fitt WK, Schmidt GW. Damage to photosystem II in symbiotic dinoflagellates: A determinant of coral bleaching. Proc Natl Acad Sci USA 1999; 96:8007-12.
69. Jones RJ, Hoegh-Guldberg O, Larkum AWD et al. Temperature induced bleaching of corals begins with impairment of the CO_2 fixation mechanism in zooxanthellae. Plant Cell Environ 1998; 21:1219-30.
70. Sawyer SJ, Muscatine L. Cellular mechanisms underlying temperature-induced bleaching in the tropical sea anemone Aiptasia pulchella J Exp Biol 2001; 204:3443-3456.
71. Douglas AE. Coral bleaching - How and why? Mar Pollut Bull 2003; 46:385-392.
72. Trench RK. Microalgal-invertebrate symbioses: A review. Endocyt Cell Res 1993; 9:135-175.
73. Perez SF, Cook CB, Brooks WR. The role of symbiotic dinoflagellates in the temperature-induced bleaching response of the subtropical sea anemone Aiptasia pallida. J Exp Mar Biol Ecol 2001; 256:1-14.
74. Franklin DJ, Hoegh-Guldberg O, Jones RJ et al. Cell death and degeneration in the symbiotic dinoflagellates of the coral Stylophora pistillata during bleaching. Mar Ecol Prog Ser 2004; 272:117-130.
75. Bibby BT, Dodge JD. Ultrastructure and cytochemistry of microbodies in dinoflagellates. Planta 1973; 112:7-16.
76. Trench RK. Nutritional potentials in Zoanthus sociathus (Coelenterata, Anthozoa). Helgol Wiss Meeresunters 1974; 26:174-216.
77. Trench RK. The biology of dinoflagellates. In: Taylor FJR, ed. Dinoflagellates in Nonparasitic Symbioses. Oxford: Blackwell, 1985:530-570.
78. Lesser MP, Farrell J. Solar radiation increases the damage to both host tissues and algal symbionts of corals exposed to thermal stress. Coral Reefs 2004; 23:367-77.
79. Gutteridge JMC, Halliwell B. Free radicals and antioxidants in the year 2000 - A historical look to the future. Ann NY Acad Sci 2000; 899:136-147.
80. Imlay JA. Pathways of oxidative damage. Ann Rev Microbiol 2003; 57:395-418.
81. Lane DP. Cell immortalization and transformation by the p53-gene. Nature 1984; 312:596-597.
82. Costas E, Aguilera A, Gonzalez-Gil S et al. Contact inhibition - Also a control for cell-proliferation in unicellular algae. Biol Bull 1993; 184:1-5.
83. Buddemeier RW, Fautin DG. Coral bleaching as an adaptive mechanism: A testable hypothesis. Bioscience 1993; 43:320-326.
84. Rowan R. Diversity and ecology of zooxanthellae on coral reefs. J Phycol 1998; 34:407-417.
85. Savage AM, Goodson MS, Visram S et al. Molecular diversity of symbiotic algae at the latitudinal margins of their distribution: Dinoflagellates of the genus Symbiodinium in corals and sea anemones. Marine Ecol Prog Ser 2002; 244:17-26.
86. Vierstra RD. Proteolysis in plants: Mechanisms and functions. Plant Mol Biol 1996; (1-2):275-302.
87. Abraham MC, Shaham S. Death without caspases, caspases without death. Trends Cell Biol 2004; 14:184-193.
88. Thornberry NA. Caspases: A decade of death research. Cell Death Diff 1999; 6:1023-1027.
89. Uren AG, O'Rourke K, Aravind LA et al. Identification of paracaspases and metacaspases: Two ancient families of caspase-like proteins, one of which plays a key role in MALT lymphoma. Mol Cell 2000; 6:961-967.
90. Villalobo E, Moch C, Fryd-Versavel G et al. Cysteine proteases and cell differentiation: Excystment of the ciliated protist Sterkiella histriomuscorum. Eukaryot Cell 2003; 2:1234-1245.
91. Vardi A, Schatz D, Beeri K et al. Dinoflagellate-cyanobacterium communication may determine the composition of phytoplankton assemblage in a mesotrophic lake. Curr Biol 2002; 12:1767-1772.
92. Okamoto OK, Hastings JW. Genome-wide analysis of redox regulated genes in a dinoflagellate. Gene 2003; 321:73-81.
93. Battinelli E, Loscalzo J. Nitric oxide induces apoptosis in megakaryocytic cell lines. Blood 2000; 95:3451-3459.

94. Mur LAJ, Carver TLW, Prats E. NO way to live; the various roles of nitric oxide in plant-pathogen interactions. J Exp Bot 2006; 57:489-505.

95. Vacca RA, Valenti D, Bobba A et al. Cytochrome c is released in a reactive oxygen species-dependent manner and is degraded via caspase-like proteases in tobacco bright-yellow 2 cells en route to heat shock-induced cell death. Plant Physiol 2006; 141:208-219.

96. Segovia M, Berges JA. Effect of inhibitors of protein synthesis and DNA replication on the induction of proteolytic activities, caspase-like activities and cell death in the unicellular chlorophyte Dunaliella tertiolecta. Eur J Phycol 2005; 40:21-30.

97. Tai V, Lawrence JE, Lang AS et al. Characterization of HaRNAV, a single-stranded RNA virus causing lysis of Heterosigma akashiwo (Raphidophyceae). J Phycol 2003; 39:343-352.

98. Nagasaki K, Shirai Y, Takao Y et al. Comparison of genome sequences of single-stranded RNA viruses infecting the bivalve-killing dinoflagellate Heterocapsa circularisquama. Appl Env Microbiol 2005; 71:8888-8894.

99. Wilson WH, Schroeder DC, Allen MJ et al. Complete genome sequencing and lytic phase transcription profile of a coccolithovirus. Science 2005; 309:1090-1092.

100. Margulis L. Symbiosis in cell evolution: Life and its environment on the early earth. San Francisco: WH Freeman, 1981.

101. Kroemer G. Mitochondrial implication in apoptosis. Towards an endosymbiont hypothesis of apoptosis evolution. Cell Death Differ 1997; 4:443-456.

102. Huettenbrenner S, Maier S, Christina Leisser C et al. The evolution of cell death programs as prerequisites of multicellularity. Mut Res 2003; 543:235-249.

103. Ameisen JC. When cells die. In: Lockshin R, Zakeri Z, Tilly J, eds. The Evolutionary Origin and Role of Programmed Cell Death in Single Celled Organisms: A new view of executioners, mitochondria, host-pathogen interactions, and the role of death in the process of natural selection. New York: Wiley-Liss, 1998:3-56.

104. Falkowski PG, Katz ME, Knoll AH et al. The evolution of modern eukaryotic phytoplankton. Science 2004; 305:354-360.

105. Knoll A. The geological consequences of evolution. Geobiology 2003; 1:3-15.

106. Delwiche CF, Kuhsel M, Palmer JD. Phylogenetic analysis of tufA sequences indicates a cyanobacterial origin of all plastids. Mol Phylogenet Evol 1995; 4:110-128.

107. Delwiche CF. Tracing the thread of plastid diversity through the tapestry of life. Am Nat 1999; 154:164-177.

108. Jain R, Rivera MC, Lake JA. Horizontal gene transfer among genomes: The complexity hypothesis. Proc Natl Acad Sci USA 1999; 96:3801-3806.

CHAPTER 12

Programmed Nuclear Death and Other Apoptotic-Like Phenomena in Ciliated Protozoa

Ana Martín González,* Silvia Díaz, Andrea Gallego and Juan C. Gutiérrez

Abstract

One of the more usual hallmarks of programmed cell death (PCD) in multicellular organisms is the nuclear chromatin condensation and the DNA fragmentation in multiple oligonucleosome length fragments. In *Tetrahymena thermophila* and other free-living ciliated protozoa, a controlled nuclear degradation process develops during conjugation, the sexual phase of the life cycle, which includes DNA condensation and later degradation, with or without previous nuclear fragmentation. The main objective of this programmed nuclear death (PND) is to remove the old macronucleus whereas a new recombinant vegetative nucleus is developing in each conjugating cell. Alternatively, in other ciliates, mainly inhabitants of terrestrial ecosystems, which have not conjugation, PND is restricted to encystment, a cellular differentiation process induced by different environmental stressors, mainly starvation, the same inducer of conjugation. The mechanism of PND is still not elucidated, but we know that it involves caspase-like proteins, an intense acid phosphatase activity and an autophagic process. Also, several reports indicate the existence of mitochondrial participation in macronuclear degradation. After analysing the updated information relative to programmed nuclear death in ciliated protozoa, we conclude that these eukaryotic microorganisms represent an alternative and good option to study the PCD process in unicellular organisms. The recently completed macronuclear genome sequencing in the model ciliate *T. thermophila*, provides insights to analyze and understand the molecular mechanisms of PCD in ciliates and, likewise, it will give rise to useful information about the origin and evolution of cell death in biological systems.

Introduction: Ciliate Organization Indicate That They Are Singular Eukaryotic Microorganisms

Ciliated protozoa are eukaryotic microorganisms with unique cellular complexity included into the subphylum Ciliophora of the phylum Alveolata.[1] At present, more than 8000 species have been described; most of them are free-living with a cosmopolitan distribution in soil and aquatic (freshwater and marine) ecosystems, where they are important grazers and contribute significantly to the regeneration of nutrients.[2] Not only their morphological structure is quite complex, but their physiological, genetical, ecological and behavioral traits may be also considered as specialized and complex. At least, there are four main singular features in ciliates:

*Corresponding Author: Ana Martín González—Departamento de Microbiología-III, Facultad de Biología, C/. José Antonio Novais, 2. Universidad Complutense (UCM), 28040-Madrid, Spain. Email: anamarti@bio.ucm.es

Programmed Cell Death in Protozoa, edited by José Manuel Pérez Martín.

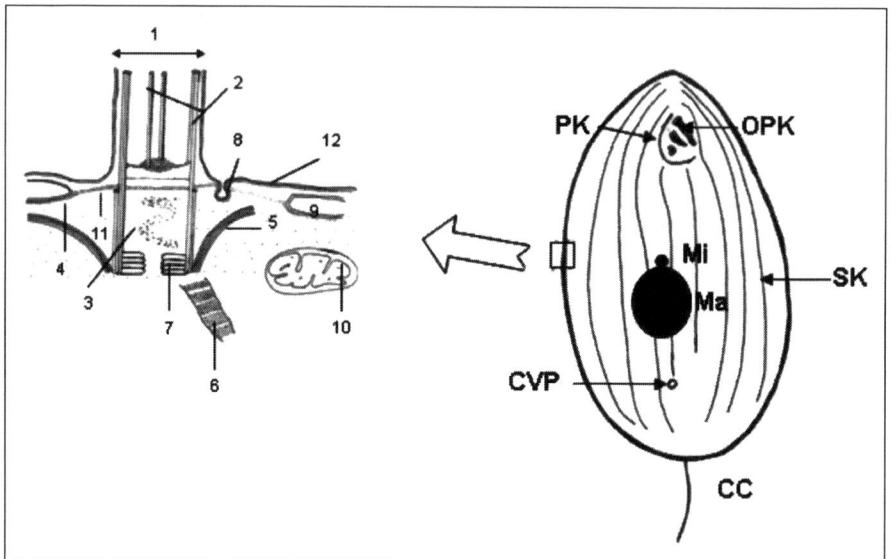

Figure 1. Diagrammatic scheme of the ventral side of an idealized ciliate (right) and a generalized representation of a portion of ciliate somatic cortex (left). Ma) macronucleus, Mi) micronucleus, CC) caudal cilium, CVP) contractile vacuole pore, OPK) oral polykinety, PK) paroral kinety, SK) somatic kinety. 1) cilium, 2) microtubules, 3) basal body or kinetosome, 4) postciliary fiber, 5) transverse fiber, 6) kinetodesmal fiber, 7) cartwheel, 8) parasomal sac, 9) alveolus, 10) mitochondria, 11) epiplasm, 12) plasma membrane.

(1) A cortical complex organization, (2) Specialized ciliature for feeding and locomotion, (3) Nuclear dualism and (4) Conjugation as sexual process. Unlike other protists (fungi, algae), ciliates have not cell wall. The cytoplasm is surrounded by an specialized portion named cortex, that is responsible for the cellular shape. Cortex (Fig. 1) is usually constituted by three components; (1) the pellicle, (2) the basal bodies or kinetosomes and associated fibrilar and microtubular elements that constitute the cellular infraciliature and (3) the extrusomes (trichocysts, toxicysts, mucocysts, etc).[3] The pellicle is formed by the cell membrane (plasmalemma), a layer of flattened vacuoles arranged in a mosaic-like pattern (alveolar layer) and the epiplasm, a proteinaceous layer underlying the alveolar vacuoles (Fig. 1). Regular invaginations of plasmalemma, named parasomal sacs, are located near to each cilia (Fig. 1) which represent sites with pinocytotic activity. Cytoplasm contains ribosomes, mitochondria, Golgi apparatus, endoplasmic reticulum, contractile fibres (myonemes), one or more contractile vacuole to control the osmotic pressure and other typical eukaryotic organelles.[4] The basic units of infraciliature are the kinetids. They are basal bodies arranged as single (monokinetids), double (dikinetids) or multiple (polykinetids) patterns with several fibrilar and microtubular associated rootless stuctures. Each kinetosome can support or not a cilium. Somatic kinetids are usually disposed in longitudinal rows (kineties) or clusters (cirri) and are involved in locomotion[3] (Fig. 1). Oral kinetids are extremely variable in their morphology and location, being involved in feeding. In general, to the left in the buccal cavity they are polykinetids and to the right margin they dispose as a single or double short row. Arrangement and composition of both, oral and somatic infraciliatures are the main morphological features used for ciliate systematic and identification (Fig. 1). Most of ciliated protozoa possess specialized cytoplasmic organelles, among them we can stand out those denominated extrusomes that can be stimulated to discharge material to the surface of the cell, thus they are exocytotic organelles involved in defense and/or depredation.[5] Although other protists, as dinoflagellates, which also have nuclear dualism, in ciliates

the organization of nuclear genome is unique.[6,7] They have two different nuclei: micronucleus and macronucleus (Fig. 1). The micronucleus or germline nucleus is diploid and always lacks of nucleoli. It contains all cellular genetic information organized in rather condensed chromosomes (n = 5-150) surrounded by a double nuclear membrane with pore complexes. Micronuclei are not transcriptionally active during growth and divide by intranuclear orthomitosis.[8] In the sexual phase of the life cycle (conjugation), micronucleus undergoes also meiosis. Micronuclei are not absolutely necessary for the cell viability, in fact there are some viable amicronucleated strains in ciliates. On the contrary, the size and stucture of the macronucleus is very different from those found in other eukaryotic cells. Macronuclei are the somatic nuclei and they are essential for cell viability. Ciliate macronucleus is considered as ampliploid, it contains only 10-90% of total genetic information which is organized in numerous acentromeric chromosomic segments (subchromosomes or minichromosomes). Some of these sequences are highly amplified, 45-90,000 copies/macronucleus. It is transcriptionally active and determines cell phenotype. Macronucleus divides by amitosis and undergoes a degeneration process during conjugation. Depending on ciliate species, one or more micronucleus and macronucleus may exist per cell.[9]

In general, food uptake in ciliates is via the cytostome. Potential food are bacteria, yeasts, microalgae and other protozoa. These particulated nutrients are trapped by phagocytosis using the oral ciliature (filter-feeders), certain specialized cytopharyngeal structures (herbivorous) or a complex cytostome-cytopharynx region (carnivorous). In any case, several food vacuoles are formed and after undergoing a digestive cycle, debris are egested by a cytoproct.[5]

A Complex Life Cycle with Diverse Alternatives

All ciliate species present a growth-division cycle in which the vegetative cell undergoes transversal bipartition, giving rise to two daughter cells, an anterior one (proter) and a posterior one (opisthe) (Fig. 2). During asexual division, new oral and somatic ciliary structures are formed according a specific pattern, micronucleus divides by mitosis and macronucleus by amitosis. Certain ciliated protozoa under unfavourable environmental conditions, such as; salinity increase, temperature shock, oxygen deprivation, environmental metabolic waste accumulation, high cellular population and, specially, starvation, enter into an alternative cycle, named encystment-excystment cycle (Fig. 2). During encystment, vegetative cell differentiates to a resting cyst.[10] Resting cyst is an ametabolic and dehydrated cell stage where the cell is surrounded by a cyst wall made up of 1-4 layers, which is an important element contributing to the high cyst resistance to extreme conditions.[11] During encystment, vegetative cell undergoes several drastic changes, including a high cell dehydration, controlled autophagy of several cellular organella, macronuclear fusion and/or chromatin condensation, reserve material formation and in many cases, partial or total resorption of both, oral and somatic cilia and infraciliatures.[12-14] When nutrient deprivation or other unfavorable environmental factor is restored to an appropiate level, excystment takes place and the resting cyst differentiates to a vegetative cell which goes to the growth-division cycle again (Fig. 2).

Besides, some ciliate species include in their biological cycle a sexual process (Fig. 2). Conjugation is the main sexual or fertilization process reported in ciliated protozoa. In general, two preconjugants of different, but complementary, mating types interact and later fuse, giving rise to a pair of conjugants.[15] Vegetative cell differentiation to a preconjugant cell is induced by starvation. Preconjugant interaction is mediated by sexual feromones (gamones), that are usually diffusible mating-type substances, although in some genera, such as *Paramecium*, they are located in the ciliary membrane.[16] During conjugation, cell pair undergoes a series of developmental processes: (1) Micronuclear meiosis, (2) Production of one or two haploid pronuclei in each conjugant, (3) Transfer of one pronucleus of each conjugant to the other, (4) Fusion of two pronuclei into a zygotic nucleus (synkaryon), (5) One or more mitotic divisions of synkarya, (6) Differentiation of these mitotic products in new micro-/macronuclei, (7) Removal of the old macronucleus, (8) Reorganization of cortical structures and (9) Separation or complete fusion of conjugants.[17] It must be pointed out that, genetic system of each exconjugant is rather different that those from initial

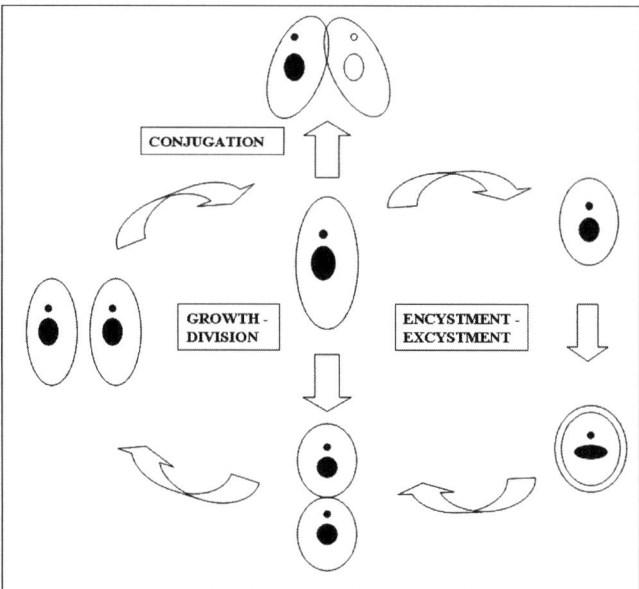

Figure 2. A schematic representation of an ideal ciliate life cycle. Three different aspects may be included in it; the Growth-Division cycle, an Encystment-Excystment cycle and sexual interaction or conjugation process.

vegetative cells, because an interchange and fusion of nuclei from both conjugant cells takes place. Although several species of ciliates present conjugation and encystment-excystment cycle in their life cycles, both phenomena are alternative and in these ciliates, a strain can conjugate and other one encyst in response to starvation, when the other conditions are appropiate.

Nuclear Phenomena during Conjugation of *Tetrahymena Thermophila*

The ciliate *Tetrahymena thermophila* is an eukaryotic microorganism model for cellular and molecular studies of protists.[18] Besides, at present, it is the only ciliate species that its macronuclear genome is completely sequenced.[19] The nuclear system is composed by one macronucleus and one diploid micronucleus, which contains five pairs of chromosomes. In this case, conjugation is temporal (conjugants separate) and isogamontic (there are not differences in size and morphological features between both conjugants) and it can be induced by prolongued starvation. Nuclear phenomena during conjugation are well-known[20,21] (Fig. 3). Soon after preconjugants (Fig. 3, steps 1-3) interact and unite in pairs, nuclear phenomena begin and continue more or less syncronically in both conjugants. So, each micronucleus separates from the macronucleus and moves to the anterior third of the cell. Then, it undergoes meiosis to generate four haploid nuclei (Fig. 3, step-4). Three of these meiotic products migrate to the posterior third of each cell where they are destroyed and disappear (Fig. 3, steps 4-6) and the remaining one, divides mitotically to yield two genetically identical gametic pronuclei (Fig. 3, step-5); a stationary pronucleus and a migratory pronucleus. After genetic exchange of migratory pronuclei between conjugating partners (Fig. 3, step-6), fusion of pronuclei with different origin takes place, giving rise to a diploid zygotic nucleus (synkaryon) (Fig. 3, step-7). These zygotic nuclei have identical genotype and undergo two sucessive mitotic divisions (postgamic or postzygotic) (Fig. 3, step-8). From the four mitotic products, one degenerates and it is resorbed, another one become micronucleus and each of the remaining two nuclei (macronuclear anlagen), differentiate to macronucleus (Fig. 3, steps 9-10).

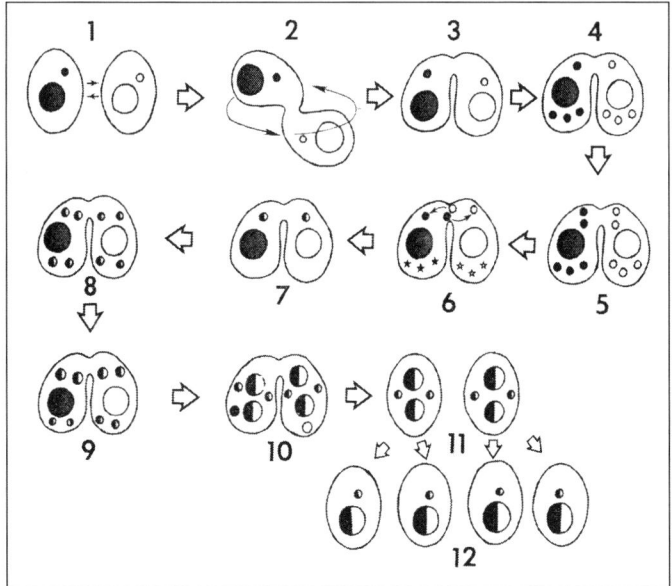

Figure 3. Schematic representation of nuclear events of *T. thermophila* conjugation. Steps 1-3) Preconjugant interactions and pair formation. Step-4) Micronuclear meiosis and selection. Step-5) Prezygotic mitosis and pronuclear differentiation. Step-6) Pronuclear exchange. Step-7) Pronuclear fusion. Step-8) First and second postzygotic mitosis. Step-9) Macronuclear anlagen. Step-10) Macronuclear anlagen and macronuclear apoptosis. Step-11) Pair separation (exconjugants) and micronuclear degradation. Step-12) Micronuclear mitosis and cell division.

Meanwhile the mitotic postgamic divisions, the old macronucleus in each conjugant condenses, becomes picnotic and finally is resorbed (Fig. 3, step-10). Therefore, each exconjugant has two macronucleus and one micronucleus. When it undergoes bipartition, micronucleus divides by mitosis and each macronucleus is segregated to a daughter cell (Fig. 3, steps 11-12).

Programmed Nuclear Death during Conjugation in *Tetrahymena*: An Apoptotic-Like Process?

The ultrastructure of the old macronucleus in each conjugant, during the degeneration process, presents several similarities to mammalian cells undergoing an apoptotic process. By this reason, some authors have suggested that this selective nuclear destruction or programmed nuclear death (PND) might be considered as a apoptotic-like process.[21,22] From the excellent transmission electron microscopy study by Weiske-Benner and Eckert,[23] it is possible to conclude that the fine structure of old (parental) macronuclei do not change significantly until the postzygotic phase of conjugation. As soon as the two most anterior nuclei from the second mitotic division of synkarya begin to become macronuclear anlagen, each old macronucleus decrease in size and become very compact and electron dense.[20] This process, which includes a strong chromatin condensation, is usually known as nuclear picnosis.[23] After this condensation stage, the old macronuclei undergo a progressive degeneration process and later they disappear by autolysis. In fact, each degenerating nuclei is included into an autophagosome membrane and then this autophagosome develops to autolysosome because a large number of lysosome-like vesicules contact and fuse with the autophagosome membrane. When most of cellular pairs have separated, the exconjugant cytoplasm only present small residual bodies in which is not possible to detect any material from nuclear origin.[23] As it occurs in apoptotic processes of pluricellular organisms, macronuclear degeneration in

T. thermophila is a highly regulated phenomenum with sucessive and fixed stages: (1) Nuclear size reduction, (2) Chromatin condensation and (3) Digestion of both DNA and protein material by macroautophagy (autophagy) until the old macronucleus disappears.

One of the most specific features of the apoptosis or programmed cell death Type I in mammals[24] is the condensation and fragmentation of the nuclei. This process is usually accompanied by an extensive degradation of chromosomal DNA into large (50-300 kb) fragments (high molecular weight DNA fragments), that are subsequently degraded into smaller fragments of oligonucleosomal size (DNA ladder).[25] Different authors consider that only the high molecular weight DNA degradation is essential for cell death, because there are several mammalian cell types that never produce DNA ladders after treatments inducing the typical apoptosis nuclear and cytoplasmic changes.[26-28] Degradation of chromosomal DNA is a well-conserved apoptotic process, which has been described in mammals, *Drosophila, Caenorhabditis elegans* and plants.[29] According to Davis et al,[22] by using total DNA extraction and agarose gel electrophoresis (1%), it is possible to detect a ladder of small DNA fragments in *Tetrahymena* samples taken from 10 h after preconjugant mixing. These fragments were sized as 175-560 ± 9-17 bp. The peak of appearance of oligonucleosome-like fragments is from 12 to 16 h, before the old macronuclei dissapear. This type of degradation has been only detected in old macronuclei, despite DNA breakage in anlagen and elimination of selected micronuclei (three of the haploid nuclei) also occurring in late conjugation.[22] Preliminary experiments from crosses of NULLI 3 mutants (micronuclear chromosome 3 nullisomic strains) with wild-type strains, indicate that while many events in conjugation are preprogrammed by the parental genome, the expression of a few new products is essential to the completion of conjugation.[22] So, in NULLI 3 conjugants, old macronuclear pycnosis and oligonucleosomal fragmentation occur normally, but the resorption of degenerating macronucleus fails.[22] Posterior to this pioneer study,[22] which support that the nuclear death (elimination of old macronucleus) in *T. thermophila* conjugation is a normal and programmed process quite similar to apoptosis nuclear process of higher organisms, other studies about PND have been focused to analyze more accurately its two main steps: chromatin condensation and DNA degradation in oligonucleosome-sized fragments. In the absence of an universal specific molecular apoptosis tracer, the TUNEL assay is the classical and most frequently used technique to identify apoptotic cells. This method detects apoptosis via DNA fragmentation and makes evident both, single- and double-stranded breaks and also early stages of apoptosis.[30] It is considered very specific, although it has been reported that several factors produce false-positive signals, therefore a confirmation by morphological examination is recommended.[31] In *T. thermophila*, later detailed studies about the cronological series of events during old macronuclei degeneration have shown that some DNA digestion is necessary and it occurs before the chromatin condensation.[32] This conclusion is supported, at least, for two different kinds of data; a comparision between the agarose gel electrophoresis patterns and cytological (DAPI nuclear stain) events during conjugation, which indicates that DNA fragmentation begins before the macronuclear condensation, although DNA ladder is more evident when macronuclear condensation has finished. Second, treatment with aurine, a structural analog of the general nuclease inhibitor aurintricarboxylic acid,[33] blocks drastically both processes; DNA condensation and DNA digestion. Similar results were obtained by exposure to ciclohexymide and actinomycin D, two inhibitors of gene expression at different levels. Finally, results from TUNEL assay demostrate that condensed macronuclei are always TUNEL-positive and even in the stage just prior to condensation, macronuclei were not fluorescent.[32] Therefore, as Nanney[34] stated may years ago, ciliates, at least *Tetrahymena*, can distinguish and recognize between two macronuclei coexisting in the same cytoplasm; the developing macronuclear anlagen and the degenerating old macronucleus. DNA fragmentation has been reported in the programmed cell death of diverse parasitic protozoa, included in the genera *Trypanosoma*,[35,36] *Leishmania*,[37,38] *Blastocystis*[39,40] and *Plasmodium*.[41,42] By other hand, programmed nuclear death is a rather unusual phenomena, although some cases have been described in fungi[43-45] as well as in invertebrates and it is related to sexual stage or fertilization in the biological cycle.[46] In mammals, at least two redundant pathways may lead to the chromatin processing during apoptosis; one of them involves Apaf-1 (apoptotic protease-activating factor 1), caspases and CAD (caspase-activated Dnase) and leads to the oligonucleosomal DNA fragmentation and advanced chromatin condensation. The other pathway, which is caspase independent, involves AIF (apoptosis inducing factor) and leads to large-scale DNA fragmentation and peripherical chromatin condensation.[47] A recent report indicate that caspase-3 is involved in the cleavage of nuclear matrix, directly or by activating other proteases.[48]

Different Involvement of Caspase-Like Activities in Programmed Nuclear Death

The caspases are a family of cysteine proteases that share an exquisite specificity for cleaving target proteins at sites next to aspartic acid residues. Caspases are the main executioner proteins of apoptosis, although this process can be executed by caspase-independent death effectors and they are also involved in diverse differentiation phenomena, inflammation, activation of pro-cytokine, T and B-cells proliferation, cell cycle control, DNA repair, etc.[49-51] No true caspases have been identified outside the metazoan clade, but caspase-like and their distant relatives, meta- and para-caspases, have been found in phylogenetically distant nonmetazoan biological groups, including plants, fungi and prokaryotes.[50,52]

As old macronuclei degeneration in ciliate conjugation is a programmed event with some similarities with apoptotic processes in higher organisms, an important point is to elucidate the participation of caspase-like activity in programmed nuclear death during conjugation of *Tetrahymena thermophila*. Two different approaches have been used, almost simultaneously, to elucidate this point. By one hand, three different caspase inhibitors have been applied and caspase-3 or caspase 3-like activity was evaluated by spectrophotometry, considering macronuclear compaction as an indicator of PND.[53] In *Tetrahymena* old macronuclear chromatin condensation during conjugation begins at about 6.5 h and, from 9 to 12 h, it is present in a greater part of cell pairs (87%). The addition of the general caspase inhibitor zVAD-fmk (Benzyloxyconnyl-Val-Ala-Asp-fluoromethylketone) to 6 h pairs, produce a drastic decrease of 12 h paired cells with macronuclear condensation, which is higher after applying increasing concentrations of ZVAD-fmk. When conjugants were exposed to two different and more specific inhibitors, such as; YVAD-CHO (N-acetyl-Tyr-Val-Ala-Asp-aldehide) for caspase-1 and, DEVD-CHO (N-acetyl-Asp-Glu-Val-Asp-aldehyde) for caspase-3, both at 0.5 mM, the results were quite similar; an inhibition of macronuclear condensation was detected, which was higher if the caspase-1 is blocked (30% of macronuclear compaction).[53] Besides, the caspase-like activities measured spectrophotometrically by the hydrolysis of specific coloured sustrates (YVAD-pNA (N-acetyl-Tyr-Val-Ala-Asp-p-Nitroaniline, for caspase-1 and DEVD-pNA (N-acetyl-Asp-Glu-Val-Asp-p-Nitroaniline, for caspase-3) in cell extracts from *Tetrahymena* 9 h conjugating pairs, gave a reduction of about 40% and 50% respectively, with regard controls, in pairs pretreated with one the above mentioned inhibitors. Measurement of caspase-like activity at different times during *Tetrahymena* conjugation, using the same colorimetric assay, showed that this activity can be detected in any time of conjugation. By means of the specific fluorescent subtrate [PhiPhiLux (OncoImmunin)] for caspase-3 and posterior analysis with fluorescence microscopy, a caspase-like activity was located into several cytoplasmic vesicles. Observations supports that caspase-3 does not play a direct role in the degradation of the old macronuclei, since nuclear differentiation is also blocked by inhibitors. It is suggested that both processes, nuclear differentiation and macronuclear degradation depend or are controlled by caspases.[53]

By other hand, the japanese group directed by Dr. Endoh also found[54] a PND process during conjugation of *Tetrahymena thermophila*, but probably due to use different ciliate strain and experimental conditions, the timing of diverse nuclear stages show important differences with regard previous reports.[22,32] So, according these authors,[54] ladder DNA fragmentation began 8 h after induction of conjugation and DNA degradation was completed around 18 h after cellular mixing. Likewise, old macronuclear condensation and DNA degradation occur almost simultaneously.[54] Incubation of *Tetrahymena* cell extracts, from 16 h pairs, with different caspase substrates (Ac-DEVD-pNA, Ac-IETD-pNA and Ac-LEHD-pNA) which are known to be specifically cleaved by caspase-3, -8 and -9, respectively, showed caspase-like activity with all substrates.[54] Increases of absorbance were in a time-dependent way. To evaluate variations in caspase-like activity during conjugation, cell extracts were prepared every 2 h. These series of cystein protease determinations showed that, caspases -8 and -9 activities increase remarkably during late conjugation stages (14-22 h), whereas caspase-3 activity was low and hardly changed during this period. Experimental evidence of no caspase-3-like activity in old macronucleus was confirmed using the caspase-3 inhibitor (Ac-DEVD-CHO). In both cases, with or without the inhibitor, Ac-DEVD-pNA assay produced similar results, existing only basal levels of caspase-3.[54] By the contrary, the activity obtained in the Ac-IETD-pNA and AcLEHD-pNA assays increased significantly during the macronuclear degeneration stage, suggesting that caspase-8 and caspase-9-like activities might be associated with nuclear death, but caspase-3-like might not be involved in it. Exploring the real selectivity of the above inhibitors in *Tetrahymena*, these authors[54] observed that when caspase-8 inhibitor (Ac-LEHD-CHO) was incubated with Ac-LEHD-pNA, no significant effect was detected. In contrast, when caspase-9 inhibitor (Ac-LEHD-CHO) was incubated with Ac-IETD-pNA, a significant inhibition was observed in *Tetrahymena* cell extracts derived from 14 and 16 h conjuganting pairs. These data can be explained by two alternative ways; first, the used inhibitors are not completely selective and second, a cascade reaction from caspase-8 to caspase-9 in *Tetrahymena* might exist. Like in *Tetrahymena*, there is indirect evidences (inhibitor effects, colorimetric measurements of caspase-like activitities, antibodies treatment, etc) for the existence of caspase-like proteases in other protozoa.[55-57] As far as we know, caspases are not found till now in unicellular organisms. However, in protozoa, distant caspase homologues, named metacaspases, were identified in *Trypanosoma brucei*, *Trypanosoma cruzi*, *Plasmodium falciparum* and *Leishmania* major.[56,58]

Autophagosome Formation and Lysosomal Enzymes Participate in Old Macronuclei Elimination

Diverse studies suggest that, although PND in *Tetrahymena* is mediated by an apoptotic-like mechanism, in which caspases -8 and -9 and probably caspase-1 participate, the final stages of old macronuclei elimination might be achieved by an autophagy (macroautophagy) process.[59-62] In ultrastructural studies, the initial autophagosome formation has not been observed in detail,[23] probably it was due to that the duration this preliminary stage is very short.[63,64] However, later stages of macronuclear autophagic process have been reported including an autophagosome stage with the inner membrane completely dissolved, corresponding to the type II defined by Nilsson,[63] and a later stage in which the autophagosome contains the old macronucleus and it is surrounded by diverse double-membraned vesicles (lysosomes) and, therefore, it is an autophagolysosome.[23] A vital double staining with the fluorophores, acridine orange (AO), which is considered as an pH indicator in living cells since it is trapped in celular acidic compartments as lysosomes and autophagolysosomes and Hoechst 33342 (HO), a DNA stain which fluoresce in blue but changes to yellow-orange in apoptotic nuclei, provide a simple method to correlate DNA degradation and acidification. Results after application this double vital stain show that lysosomes in conjuganting cells are preferentially located in the posterior part of the cells around the degenerating old macronucleus (TUNEL positive), whereas in vegetative cells, lysosomes are randomly distributed throughout the cytoplasm.[59] Besides, this methodology also differentiates between early and late stages of macronuclear degeneration; the yellow color is associated with early apoptotic nuclei and the red fluorescence would indicate later stages. This differential staining migth be due to a decrease in the pH of the degenerating macronucleus, corroborating the idea that lysosomes might play an important role during PND.[59]

Acid phosphatase is the main lysosomal acid hydrolase. In ciliated protozoa, acid phosphatase had been applied as a marker of lysosomes, digestive vacuoles and phagolysosomes.[65-67] It has been reported that in *Tetrahymena thermophila*, autolysosomes contain about 80% of acid phosphatase and starvation conditions induce a differential increase of this hydrolase activity, although other important lysosomal enzymes remained essentially unaltered.[66,68] Being in mind these considerations, a monitorization of acid phosphatase activity during *Tetrahymena* conjugation was carried out using a colorimetric assay.[60] In contrast with the author's hypothesis, repeated experiments conclude that there is no significant change in the amount of acid phosphatase activity during conjugation, associated with macronuclear degradation.[60] However, these data are quite different from those obtained in experiments concerned to cytochemical acid phosphatase activity localization in relation with nuclear events during conjugation, which have reported an intimate association between lysosomes and the degrading macronuclei. The condensed degenerating macronucleus becomes acidic containing acid phosphatase, probably due to the fusion of lysosomal content with the nuclear autophagosome to form an autophagolysosome. The colocalization of acid phosphatase with DAPI positive material supports the hypothesis that lysosomal material and chromatin become enclosed in a common compartment. According to the authors,[60] autophagy mediated by the direct lysosomal way would be involved in the PND of *Tetrahymena*. A two-stage process is proposed for the elimination of the old macronucleus in this ciliate; an apoptotic induction followed by apoptotic completion. First stage was evidenced by TUNEL staining,[32] DNA digestion into nucleosome-sized fragments,[22,54] a caspase-like activity increasing during macronuclear elimination[53,54,61] and the blockage of nuclear condensation by inhibitors of caspase activity.[53,54,61] Second stage or autophagy was corroborated by the acidification of condensed macronucleus and the location of acid phosphatase activity in degrading macronucleus.[60] In this proposed model, the lamellar vesicles surrounding the condensed old macronucleus, observed by electron microscopy,[23] would be fused to give rise an autophagosomal membrane around old macronucleus.

Mitochondria Might Be also Involved in *Tetrahymena's* PND

Alternatively, there are some evidences that support an important role of mitochondria in programmed nuclear death of *Tetrahymena thermophila*. Using the DePsipher dye, it is possible to detect the loss of mitochondrial membrane potential, because this dye accumulates in multimeric form in the mitochondrial intermembrane spaces and fluoresces in red when mitochondria retain membrane potential. On the other hand, the dye is in monomeric form and disperses throughout the cytoplasm when mitochondrial membrane is lost, emitting a green fluorescence.[61] Fluorescent mitochondria images can be correlated with nuclear phenomena made evident by DAPI staining. By this way, it was possible to detect that the mitochondrial fluorescence pattern remained unchanged during conjugation until the macronuclear degeneration began.[61] After this, the staining pattern of the precondensed old macronuclei changed drastically and the nuclei become green. This stage corresponds to autophagosome formation and the observations indicate that a certain number of mitochondria are also englobed inside autophagosome loosing their membrane potential. Similar patterns were obtained with the Green MitoTracker fluorescent stain method suggesting that some mitochondria are trapped by the autophagosome and they are digested inside.[61] By other hand, an interesting hypothesis about the possibility of releasing potential executer molecules from broken mitochondria is given in this study.[61] In fact, an endonuclease quite similar to the mammalian endonuclease G has been isolated and characterised from *T. thermophila* mitochondria.[61] Its optimal pH is 6.0-6.5 and it is sensitive to zinc ions.[61,62] Incubation of mitochondria with isolated macronuclei, under environmental conditions in which mitochondria usually burst and putative DNase is liberated, generated DNA fragments of aproximately 150-400 bp. These sizes corresponded roughly to the monomeric and dimeric forms of macronuclear DNA ladder in PND of *Tetrahymena*. According to the obtained results,[61,62] a model for DNA degradation is proposed, in which the mitochondria might be the main death executor of PND, signaling a pathway involving an endonuclease activity.[61,62] Although no experimental evidence is supplied, phosphatidylserine is postulated as a possible signal recognized by autophagosome.[62]

Finally, it must be pointed out that a novel chromodomain-containing protein (Pdd1p) has been localized simultaneously in both, developing macronucleus and degenerating old macronucleus in *Tetrahymena* conjugation.[69] Pdd1p (programmed DNA degradation protein 1) is localized in electrondense heterochromatic structures that contain germline-specific delection elements and it is also associated with parental macronucleus during the terminal stages of apoptosis.[70] This protein represents the first molecular link between heterochromatin assembly and programmed DNA degradation.

Other Cases of PND Described in Free-Living Ciliated Protozoa

Like *Tetrahymena thermophila*, other ciliate species include a conjugation process in their life cycles. In some species, such as *Paramecium caudatum*, parental (old) macronucleus of each conjugant undergoes also a PND phenomenon, but it shows some different aspects with regard *Tetrahymena* PND process. Meanwhile the synkarya are in postzygotic division stage, the old macronucleus exhibits remarkable morphological changes, that includes an elongation and subsequent fragmentation into about 40-50 pieces.[71,72] These fragments do not degenerate after differentiation of new nuclei, so the new macronucleus and the parental macronuclear fragments coexist for a prolonged period in the daugters cells.[73] The biological significance of this nuclear coexistance is still unclear. Southern blot analysis using 18S rDNA as a probe, revealed that degradation of macronuclear fragments occurred between the fifth and the sixth cell cycle of exconjugants, so in *P. caudatum*, there is a delayed degradation of parental macronuclear DNA.[74] This macronuclear behavior is markedly different from those of most ciliates, but similar to that found in other *Paramecium* species.[73] Results after staining with acridine orange suggest that autophagosomes fused with lysosomes participate in this nuclear degradation.[74]

In several free-living ciliated protozoa, specially soil ciliates, any sexual process (conjugation) has not been reported, at present, in their life cycles, but under unfavourable environmental

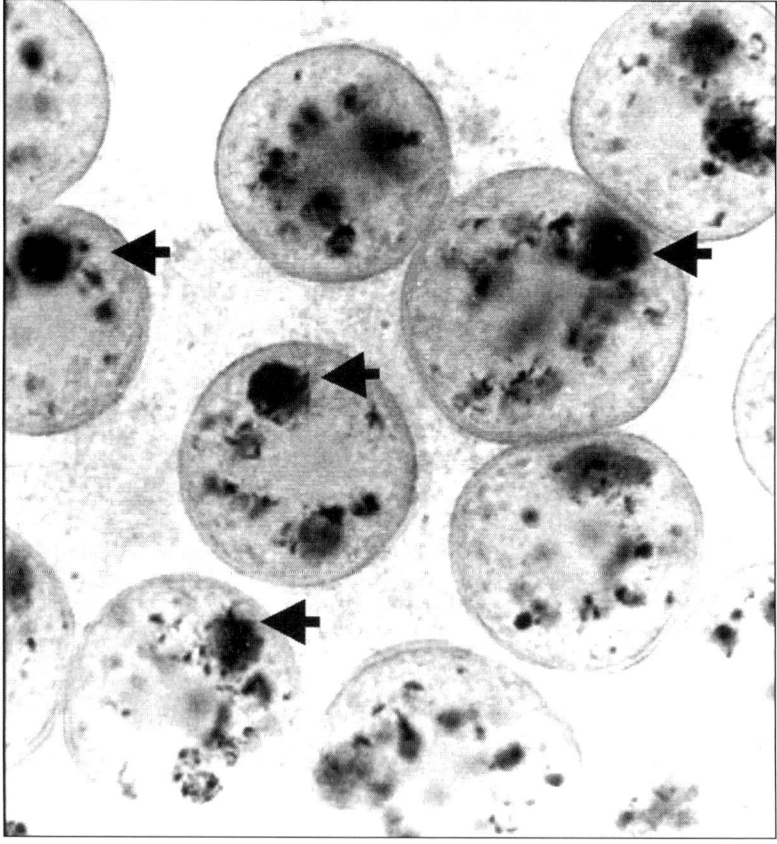

Figure 4. Light micrograph of 22 h precystic cell population of the ciliate *Colpoda inflata*, showing a positive and intense acid phosphatase activity (arrows) localized in a delimited cytoplasmic region, which correspond with macronuclear DNA extrusion body (DAPI positive). Magn.: 1,000 x.

conditions, mainly starvation or a deficiency of any essential nutrient, vegetative cells undergo drastic morphological and physiological changes to differentiate into resting cysts.[10,13] During many years, we have studied very diverse ultrastuctural, physiological and molecular aspects of encystment in colpodid ciliates, mainly the species *Colpoda inflata*. All colpodids are terrestrial microorganisms with a biological cycle including a growth-division cycle and a encystment-excystment cycle (Fig. 2). In all species of this genus, the asexual division is multiple and takes place inside a temporal (division) cyst; a vegetative cell divides, giving rise to four daugther cells. During both, asexual division and encystment, the macronucleus of each daughter cell or precystic cell undergoes a process named macronuclear extrusion which might be considered as a PND.[75,76] In this process, a portion of macronucleus (extrusion body), probably constituted by highly repeated DNA, separates and become condensed. Later, this extrusion body is degraded and disappears.[75] In a cytochemical study on the pattern changes of acid phosphatase activity during encystment of *Colpoda inflata*, we have observed that in 22-23 h precystic cells, a rounded mass with an intense acid phosphatase activity is make evident in the cytoplasm (unpublished data) (Fig. 4). This mass corresponds with the extrusion body, because it can be stained with specific DNA dyes (DAPI staining, unpublished data). Besides, at this time (22-23 h after encystment induction), the maximum expression of a macronuclear

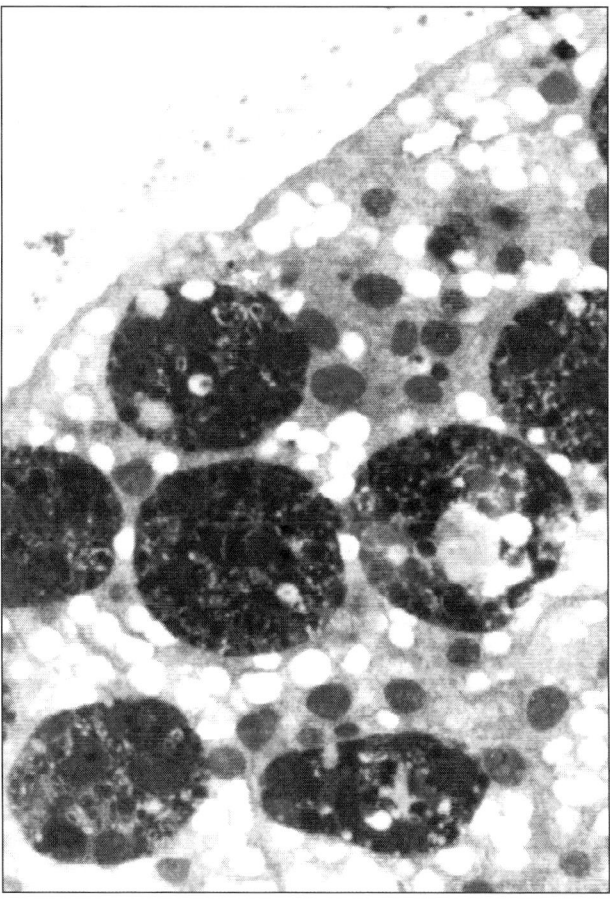

Figure 5. Transmission electron micrograph of numerous autophagosomes in the cytoplasm of a sticotrichous ciliate early precystic cell. Magn.: 10,000 x.

gene encoding an acid phosphatase (lysosomal hydrolase) is detected (unpublished data). The main environmental inducer of colpodid encystment is starvation.[10,12,13] During the early stages of this eukaryotic microbial differentiation process, an intense autophagic activity is detected and numerous autophagosomes/autolysosomes, which exhibit similar ultrastructural features to those of mammalian cells and yeasts,[64,77] are distributed throughout the ciliate precystic cytoplasm[12,13] (Fig. 5). This autophagosomes contain numerous cellular material, specially ribosomes and mitochondria.[12,13] Ciliate encystment autophagy has been associated to a protein turnover, because there are evidences of protein degradation as well as a biosynthesis of new proteins during *Colpoda* encystment.[10] Although a more intensive experimental work should be carried out, at present, all data support that autophagy is involved in the PND reported during encystment of *Colpoda inflata*.

Apoptotic-Like Processes Induced in Ciliates Support the Autophagy Implication in PND

Staurosporine is a non selective inhibitor of protein kinases, including protein kinase C, which blocks cell proliferation and also triggers a caspase-dependent induction of apoptosis in mammalian cells.[78] When vegetative cells of *Tetrahymena thermophila* were exposed to staurosporine, an inhibition of cell proliferation was detected.[79] Besides, staurosporine also induces in *Tetrahymena* a cell death process, which has similarities to both apoptosis and autophagic death, being dependent on de novo transcriptional and translational processes.[78] Staurosporine treated cells (20-30 nM) are almost motionless and presents in some cases deformations, similar to the typical apoptotic membrane blabbing. These cells form autophagosomes, some mitochondria become electrondense, the macronuclear nucleoli assemble and the macronucleus become pycnotic.[78] These ultrastructural changes are quite similar to those reported in *Tetrahymena* under prolonged starvation,[63,80] as well as those reported in several species of ciliates treated with heavy metals (Cd, Zn).[81] Therefore, certain stress environmental conditions might induce an active death process associated to autophagy.

Three different approaches have originated a series of indirect evidences suggesting that autophagy is involved in the PND during the *Tetrahymena* conjugation. By one hand, the treatment with C_2 ceramide caused rounding off in *Tetrahymena* cells, macronuclear condensation and DNA degradation. Exposure with 50 μM of C_2 ceramide resulted in the production of small DNA fragments (200-400 Kb).[82] Ceramides are products from sphingomielin breakdown and they are considered as key mediators in signaling pathways of mammals, including differentiation, growth arrest and apoptosis.[83] Ceramides are also involved in celular stress and induce macroautophagy in mammals.[84] Recent studies have demostrated that exogenous C_2 ceramide stimulates autophagy in human colon cancer cells by two non-exclusive mechanisms; stimulating the expression of a autophagy gene encoding the protein Beclin-1, involved in the first steps of autophagy[85] and interfering the class I PI3K signaling pathway by inhibition of protein kinase B.[86] According to Kihara et al,[87] most of Beclin 1 and phosphatidylinositol 3-kinase form a complex at the trans-Golgi network which control autophagy, by sorting autophagosomal components and lysosomal proteins. However, it has also been suggested that the novo biosynthesis of ceramides play a role in the induction of apoptosis.[88] Furthermore, ceramide production seems to be also linked to the induction of necrosis, because in mammalian cells cell-permeable C6- C2-ceramide induced cell death without caspase-3 activation, DNA fragmentation, cell shrinkage and DNA condensation.[89]

In mammals, phosphoinositide 3-kinase (PI 3-kinase) catalizes the formation of phosphoinositide 3-phosphate and induces autophagic programmed cell death.[90] When *T. thermophila* conjugating cells were treated with three well-known PI 3-kinase inhibitors; wortmannin (25 nM), 3-methyladenine (10 mM) or LY294002 (100 μM), several drastic changes at different stages during development were observed.[91] Although they have quite different action mechanisms; wortmannin and LY294002 bind to the catalytic subunit of PI 3-kinase, while 3-methyladenine interacts with the regulatory subunit, all inhibitors blocked the acidification of the old macronucleus and induced nuclear over-proliferation, the last one probably was due to the inhibition of degradation process involving to three of meiotic products.[92] Authors[92] conclude that the PI 3-kinase pathway is involved in conjugation PND of *T. thermophila* and it is required for acidification and degradation of the old macronuclei. However, we must consider that 3-metyladenine is a general inhibitor of phosphatidyl-inositol 3-kinases and thus it may affect to numerous cellular process besides autophagy.[93]

In the sticotrichous ciliate *Stylonychia lemnae*, around 98% of the macronuclear genome is eliminated during conjugation macronuclear differentiation; so, only 2% of the genome contain all the information required for the cellular vegetative growth.[6] In this ciliate species, a nuclear protein, named Spdd1p (*Stylonychia* programmed DNA degradation protein 1), has been described to be homologous to the above mentioned protein Pdd1p from *Tetrahymena*.[94] Interestingly, this protein seems to be involved not only in DNA degradation of both, the old macronuclei and the macronuclear anlagen development during the conjugation of *Stylonychia*, but it also participates in, a TUNEL positive, limited DNA nuclear (micronuclei and macronuclei) degradation in starved vegetative cells. How far Spdd1p play a role in these processes is not still known.[94] Controlled autophagy is one of the main phenomena induced by nutrient deprivation and it has been associated with cellular constituents recycling, in order to maintain an appropriate cellular bioenergetics level in absence of a nutrient supply. In the related species *Stylonychia mytilus*, a programmed process of autophagy has been described during conjugation. Numerous phosphatase positive autophagosomes (autolysosomes) has been observed in both conjugant and exconjugant cells, until 60 h after post-conjugant development.[95]

Concluding Remarks and Future Prospects

Diverse ciliated protozoa, including *Tetrahymena thermophila*, undergo an unique process of PND during the sexual process or conjugation. Furthermore, certain ciliates, specially those included in the order Colpodida, show a partial PND during encystment. Although evidences are still scarce, controlled autophagy and caspase-like proteins seem to be involved in ciliate PND. Some research works also support the implication of mitochondria in macronuclear death. By other hand, some morphological and physiological features of programmed cell death (Types I and II) have been observed in ciliated protozoa under stress conditions, particularly when cells are exposed to staurosporine and heavy metals. Like in other protozoa[36-39,96-98] and microalgae,[99] PND in *Tetrahymena* is induced by many different unfavorable environmental conditions. In all these eukaryotic microorganisms, PCD features are not identical to the typical pattern of cell death described in mammals, but it is quite usual to detect a mixture of morphological and physiological traits of apoptotical and autophagic processes. In some protozoa (*Giardia, Trichomonas*), these differences might be due to the absence of mitochondria.[97,98] The existence of PND in unicellular organisms, both prokaryotes and eukaryotes, support the ancient origin of apoptosis although in some cases the biological meaning of these processes remains obscure.[100] In our opinion, to use a broad spectrum of caspase inhibitors is not an appropiate approach to study cell death for three reasons: (1) they present a poor membrane permeability, (2) they are modified by peptide pseudosubstrates and (3) these inhibitors may also affect to other proteins unintentionally.[101] This last reason can also be applied to the study of authophagic cell death. By these reasons, we think that novel strategies should be applied to elucidate the mechanisms involved in the cell death of protozoa. As genome sequenciation of several protozoa is already finished or almost completed (*Trypanosoma, Giardia, Plasmodium, Leishmania, Tetrahymena, Emiliania*), a *in silico* genomic analysis of diverse putative cell death mechanisms in protozoa might make evident similarities and differences with regard cell death in other organisms, mainly yeasts and mammals. Expression studies of putative ortologues will provide valuable data to know better the cell death process in these organisms. At present, we are involved in a research project to study the cell death in *Tetrahymena thermophila*, using this last approach, because homologous sequences to important genes encoding crucial factors in yeast or mammalian apoptosis (for instance; AIF, Beclin-1, Parp-1, DED, Apaf-1, Tor, Atg 1,5,7, etc.,) seem to be present in the macronuclear genome of this ciliated protozoa.

Acknowledgements

The research work involving the study of the cell death in *Tetrahymena thermophila* is supported by the grant CGL2005-00548 from MEC. Ph.D. scholarship from Complutense University (UCM) to A.G.

References

1. Lee JJ, Leedale GF, Bradbury P, eds. Illustrated Guide to the Protozoa. 2nd ed. Blackwell Publishing, 2005.
2. Fenchel T. Ecology of Protozoa. 2nd ed. Madison: Science Tech, 1987.
3. Paulin JJ. Morphology and cytology of ciliates. In: Hausmann K, Bradbury PC, eds. Ciliates. Cells as organisms. Stuttgart: Gustav Fischer, 1996:1-40.
4. Hausmann K, Hülsmann N, Radek R. Protistology. 3rd ed. Berlin: Schweizerbart'sche Verlagbuchhandlung, 2003.
5. Radek R, Hausmann K. Phagotrophy in ciliates. In: Hausmann K, Bradbury PC, eds. Ciliates. Cells as organisms. Stuttgart: Gustav Fisher, 1996:197-219.
6. Prescott DM. The DNA of ciliated Protozoa. Microbiol Rev 1994; 58:233-266.
7. Prescott DM. Evolution of DNA organization in hypotrichous ciliates. Ann NY Acad Sci 1999; 870:301-313.
8. Raikov IB. The Protozoan Nucleus. Morphology and Evolution. Wien: Springer-Verlag, 1982.
9. Raikov IB. Nuclei of ciliates. In: Hausmann K, Bradbury PC, eds. Ciliates. Cells as organisms. Stuttgart: Gustav Fisher, 1996:221-242.
10. Gutiérrez JC, Martín-González A, Matsusaka T. Towards a generalized model of encystment (cryptobiosis) in ciliates: a review and a hypothesis. BioSystems 1990; 24:17-24.
11. Gutiérrez JC, Díaz S, Ortega R et al. Ciliate resting cyst wall: a comparative review. Recent Res Develop Microbiol 2003; 7:361-379.
12. Gutiérrez JC, Izquierdo A, Martín-González A et al. Cryptobiosis in colpodid ciliates: A microbial eukaryotic differentiation model. Recent Res Devel Microbiol 1998; 2:1-15.
13. Gutiérrez JC, Callejas S, Borniquel S et al. Ciliate cryptobiosis: a microbial strategy against environmental starvation. Int Microbiol 2001; 4:151-157.
14. Martín-González A, Benítez L, Gutiérrez JC. Ultrastructural analisis of resting cysts and encystment in Colpoda inflata. 2. Encystment process and a review of ciliate resting cyst classification. Cytobios 1992b; 72:93-103.
15. Miyake A. Physiology and biochemistry of conjugation in ciliates. In: Levandowsky M, Hutner SH, eds. Biochemistry and Physiology of Protozoa. New York: Academic Press, 1981:121-198.
16. Watanabe T. The role of ciliary surfaces in mating in Paramecium. In: Bloodgood RA, eds. Ciliary and Flagellar Membranes. New York: Plenum Press, 1990:149-171.
17. Miyake A. Fertilization and sexuality in ciliates. In: Hausmann K, Bradbury PC, eds. Ciliates. Cells as organisms. Stuttgart: Gustav Fisher, 1996:243-290.
18. Collins K, Gorovsky MA. Tetrahymena thermophila. Curr Biol 2005; 10:R317-318.
19. Turkewitz AP, Orias E, Kapler G. Functional genomics: the coming of age for Tetrahymena thermophila. Trends in Genetics 2002; 18:35-40.
20. Martindale DW, Allis CD, Bruns PJ. Conjugation in Tetrahymena thermophila. A temporal analysis of cytological stages. Exp Cell Res 1992; 140:227-236.
21. Orias E. Ciliate conjugation. In: Gall JG, ed. The Molecular Biology of Ciliated Protozoa. Academic Press, 1986:45-84.
22. Davis MC, Wart JG, Herrick G et al. Programmed nuclear death: Apoptotic-like degradation of specific nuclei in conjugating Tetrahymena. Dev Biol 1992; 154:419-442.
23. Weiske-Benner A, Eckert WA. Differentiation of nuclear structure during the sexual cycle in Tetrahymena thermophila; II. Degeneration and autolysis of macro- and micronuclei. Differentiation 1987; 34:1-12.
24. Lockshin RA, Zaheri Z. Apoptosis, autophagy and more. Int J Biochem Cell Biol 2004; 36:2405-2419.
25. Walker PR, Sikorska M. New aspects of the mechanism of DNA fragmentation in apoptosis. Biochem Cell Biol 1997; 75:287-299.
26. Boix J, Llecha N, Yuste VJ et al. Characterization of the cell death process induced by staurosporine in human neuroblastome cell lines. Neuropharmacology 1997; 36:811-821.
27. Walker PR, Leblanc J, Carson C et al. Neither caspase-3 nor DNA fragmentation factor is required for high molecular weight DNA degradation in Apoptosis. Ann NY Acad Sci 1999; 887:48-59.
28. Yuste VJ, Bayascas JR, Llecha N et al. The absence of oligonucleosomal DNA fragmentation during apoptosis of IMR-5 neuroblastoma cells. J Biol Chem 2001; 276:22323-22331.

29. Nagata S, Nagasake H, Kawane K et al. Degradation of chromosomal DNA during apoptosis. Cell Death Differ 2003; 10:108-116.
30. Negoescu A, Guillermet C, Lorimier P et al. Importance of DNA fragmentation in apoptosis with regard to TUNEL specificity. Biomed & Pharmacother 1998; 52:252-258.
31. Pulkkanen KJ, Laukkanen MO, Naarala J et al. False- positive apoptosis signal in mouse kidley and liver detected with TUNEL assay. Apoptosis 2000; 5:329-333.
32. Mpoke SS, Wolfe J. DNA digestion and chromatin condensation during nuclear death in Tetrahymena. Exp Cell Res 1996; 225:357-365.
33. Batistatou A, Greene LA. Intranucleosomal DNA cleavage and neuronal cell survival/death. J Cell Biol 1993; 122:523-532.
34. Nanney D. Nucleocytoplasmic interaction during the conjugation in Tetrahymena. Biol Bull 1953; 105:133-148.
35. Ameisen JC, Idziorek T, Brillant-Mulot O et al. Apoptosis in a unicellular eukaryote (Trypanosoma cruzi): implications for the evolutionary origin and role of programmed cell death in the control of cell proliferation, differentiation and survival. Cell Death Differ 1995; 2:285-300.
36. Welburn SC, Dale C, Ellis D et al. Apoptosis in procyclic Trypanosoma brucei rhodesiense in vitro. Cell Death Differ 1996; 3:229-236.
37. Arnoult D, Akarid K, Godet A et al. On the evolution of programmed cell death: apoptosisod the unicellular eukaryote Leishmania major involves cysteine proteinase activation and mitochondrion permeabilization. Cell Death Differ 2002; 9:65-81.
38. Lee N, Bertholet S, Debrabant A et al. Programmed cell death in the unicellular parasite Leishmania. Cell Death Differ 2002; 9:53-64.
39. Nasirudeen AMA, Tan KS, Sing M et al. Programmed cell death in a human intestinal parasite, Blastocystis hominis. Parasitology 2001a; 123:235-246.
40. Tan KSW, Nasirudeen AMA. Protozoan programmed cell death-insights from Blastocystis deathstyles. Trends Parasitol 2005; 21:547-550.
41. Al-Olayan EM, Williams GT, Hurd H. Apoptosis in the malarial protozoan, Plasmodium bergeri: a possible mechanism for limiting intensity of infection in the mosquito. Int J Parasitol 2002; 32:1133-1144.
42. Deponte M, Becker K. Plasmodium falciparum- do killers commit suicide? Trends Parasitol 2004; 20:165-169.
43. Maheshwari R. Nuclear behavior in fungal hyphae. FEMS Microbiol Lett 2005; 249:7-14.
44. Marek SM, Wu J, Glass NL et al. Nuclear DNA degradation during heterokaryon incompatibility in Neurospora crassa. Fungal Genet Biol 2003; 40:126-137.
45. Nevzglyadova O, Artyomov AV, Mikhailova EV et al. Bud selection and apoptosis-like degradation of nuclei in yeast heterokayons: a KAR1 effect. Mol Gen Genomics 2005; 274:419-427.
46. Buckland-Nicks J, Tompkins G. Paraspermatogenesis in Ceratostoma foliatum (Neogastropoda): Confirmation of programmed nuclear death. J Exp Zool 2005; 303A:723-741.
47. Susin SA, Daugas E, Ravagnan L et al. Two distinct pathways leading to nuclear apoptosis. J Exp Med 2000; 192:571-579.
48. Kivinen K, Kallajoki M, Taimen P. Caspase-3 is required in the apoptotic disintegration of the nuclear matrix. Exp. Cell Res 2005; 311:62-73.
49. Boatright KM, Salvesen GS. Mechanisms of caspase activation. Curr Op Cell Biol 2003, 15:725-731.
50. Boyce M, Degterev A, Yuan J. Caspases: an ancient cellular sword of Damocles. Cell Death Differ 2004; 11:29-37.
51. Garrido C, Kroemer G. Life's smile, death's grin: vital funtions of apoptosis-executing proteins. Curr Op Cell Biol 2004; 16:639-646.
52. Uren AG, O'Rourke K, Aravind LA et al. Identification of paracaspases and metacaspases: two ancient families of caspase-like proteins, one of which plays a key role in MALT lymphome. Mol Cell 2000; 6:961-967.
53. Ejercito M, Wolfe J. Caspase-like activity is required for programmed nuclear elimination during conjugation in Tetrahymena. J Euk Microbiol 2003; 50:427-429.
54. Kobayashi T, Endoh H. Caspase-like activity in programmed nuclear death during conjugation of Tetrahymena thermophila. Cell Death Differ 2003; 10:634-640.
55. Nasirudeen AMA, Singh M, Yap EH et al. Blastocystis hominis: evidence for caspase-3-like activity in cells undergoing programmed cell death. Parasitol Res 2001b: 87:559-565.
56. Mottram JC, Helms MJ, Coombs GH et al. Clan CD cysteine peptidases of parasitic protozoa. Trends Parasitol 2003; 19:182-187.
57. Zangger H, Mottram JC, Fasel N. Cell death in Leishmania induced by stress and differentiation: programmed cell death or necrosis? Cell Death Differ 2002; 9:1126-1139.
58. Kosec G, Alvarez V, Agüero F et al. Metacaspases in Trypanosoma cruzi: Possible candidates for programmed cell death mediators. Mol Biochem Parasitol 2006; 145:18-28.

59. Mpoke SS, Wolfe J. Differential staining of apoptotic nuclei in living cells: Application to macronuclear elimination in Tetrahymena. J Histochem Cytochem 1997; 45:675-683.
60. Lu E, Wolfe J. Lysosomal enzymes in the macronucleus of Tetrahymena during its apoptosis-like degradation. Cell Death Differ 2001; 8:289-297.
61. Kobayashi T, Endoh H. A possible role of mitochondria in the apoptotic-like programmed nuclear death of Tetrahymena thermophila. FEBS J 2005; 272:5378-5387.
62. Endoh H, Kobayashi T. Death harmony played by nucleus and mitochondria. Nuclear apoptosis during conjugation of Tetrahymena. Autophagy 2006; 2:129-131.
63. Nilsson JR. On starvation-induced autophagy in Tetrahymena. Carsberg Res Commun 1984; 49:323-340.
64. Dunn WA Jr. Studies of the mechanisms of autophagy: Formation of the autophagic vacuole. J Cell Biol 1990; 110:1923-1933.
65. Fok AK, Muraoka JH, Allen RD. Acid phosphatase in the digestive vacuoles and lysosomes of Paramecium caudatum: A timed study. J Euk Microbiol 1984; 31:216-220.
66. Kiy T, Vosskühler C, Rasmussen L et al. The three pools of lysosomal enzymes in Tetrahymena thermophila. Exp Cell Res 1993; 205:286-292.
67. Skotarczak B. The formation of primary and secondary lysosomes in Balantidium coli, Ciliata. Folia Histochem Cytobiol 1999; 37:261-265.
68. Rasmussen L, Florin-Christensen M, Florin-Christensen J et al. Differential increase in activity of acid phosphatase induced by phosphate starvation in Tetrahymena. Exp Cell Res 1992; 201:522-525.
69. Maddireddi MT, Davis MC, Allis CD. Identification of a novel polypeptide involved in the formation of DNA-containing vesicles during macronuclear development in Tetrahymena. Dev Biol 1994; 165:418-431.
70. Maddireddi MT, Coyne RS, Smothers JF et al. Pdd1p, a novel chromodomain-containing protein, links heterochromatin assembly and DNA elimination in Tetrahymena. Cell 1996; 87:75-84.
71. Mikami K, Hiwatashi K. Macronuclear regeneration and cells division in Paramecium caudatum. J Euk Microbiol 1975; 22:536-540.
72. Mikami K. Internuclear control of DNA synthesis in exconjugants cells of Paramecium caudatum. Chromosoma 1979; 73:131-142.
73. Kimura N, Mikami K. Interactions between newly developed macronuclei and maternal macronuclei in sexually immature multinucleate exconjugants of Paramecium caudatum. Differentiation 2003; 71:337-345.
74. Kimura N, Mikami K, Endoh H. Delayed degradation of parental macronuclear DNA in programmed nuclear death in Paramecium. Genesis 2004; 40:15-21.
75. Martín-González A, Benítez L, Gutiérrez JC. Cortical and nuclear events during cell division and resting cyst formation in Colpoda inflata. J Euk Microbiol 1991; 38:338-344.
76. Martín-González A, Palacios G, Gutiérrez JC. Macronuclear chromatin changes during encystment in the ciliate Colpoda inflata: Formation of cristal-like structures in the resting cyst chromatin and nucleolar condensation. Eur J Protistol 2001; 37:121-136.
77. Baba M, Takehisge K, Baba N et al. Ultrastructural analysis of the autophagic process in yeast: detection of autophagosomes and their characterization. J Cell Biol 1994; 124:903-913.
78. Christensen ST, Chemnitz J, Straarup EM et al. Staurosporine-induced cell death in Tetrahymena thermophila has mixed characteristics of both apoptotic and autophagic degeneration. Cell Biol Inter 1998; 22:591-598.
79. Straarup EM, Sshousboe P, Quie H et al. Effects of protein kinase C activators and staurosporine on cell survival, proliferation and protein kinase activity in Tetrahymena thermophila. Microbios 1997; 91:181-190.
80. Levy MR, Elliot AM. Biochemical and ultrastructural changes in Tetrahymena pyriformis during starvation. J Euk Microbiol 1968; 15:208-222.
81. Martín-González A, Borniquel S, Díaz S et al. Ultrastructural alterations in ciliated protozoa under heavy metal exposure. Cell Biol Inter 2005; 29:119-126.
82. Kovács P, Hegyesi H, Köhidai L et al. Effect of C_2 ceramide on the inositol phospholipid metabolism (uptake of ^{32}P, ^3H-serine and ^3H-palmitic acid) and apoptosis-related morphological changes in Tetrahymena. Comp Biochem Physiol C 1999; 122:215-224.
83. Hannun YA, Obeid LM. Ceramide: an intracellular signal for apoptosis. TIBS 1995; 20:73-77.
84. Klionsky DJ, Emr S. Autophagy as a regulated pathway of cellular degradation. Science 2000; 290:1717-1721.
85. Edinger AL, Thompson CR. Death by design: apoptosis, necrosis and autophagy. Curr Op Cell Biol 2004; 16:663-669.
86. Scarlatti F, Bauvy C, Ventruti A et al. Ceramide-mediated macroautophagy involves inhibition of protein kinase B and up-regulation of Beclin 1. J Biol Chem 2004; 279:18384-18391.

87. Kihara A, Kabeya Y, Ohsumi Y et al. Beclin-phosphatidylinositol 3-kinase complex fuctions at the trans-Golgi network. EMBO reports 2001; 21:330-335.
88. Hannun YA, Obeid LM. Ceramide-centric universe of lipid-mediated cell regulation: stress encounters the lipid kind. J Biol Chem 2002; 277:25847-25850.
89. Hetz CA, Torres V, Queso AFG. Beyond apoptosis: non apoptotic cell death in physiology and disease. Biochem Cell Biol 2002; 83:578-579.
90. Petiot A, Ogier-Denis E, Bloommart EF et al. Distinct classes of phosphatidylinositol 3'-kinases are involved in signaling pathways that control macroautophagy in HT-29 cells. J Biol Chem 2000; 275:992-998.
91. Yakisich JS, Kapler GM. The effect of phosphoinositide 3-kinase inhibitors on programmed nuclear degradation in Tetrahymena and the fate of surviving nuclei. Cell Death Differ 2004; 11:1146-1149.
92. Walker EH, Pacold ME, Perisic O et al. Structural determinants of phosphoinositide 3-kinase inhibition by wortmannin, LY294002, quercetin, myricetin and staurosporine. Mol Cell 2000; 6:906-919.
93. Blommaart EF, Krause U, Schellens JP et al. The phosphatidylinositol 3-kinase inhibitors wortmannin and LY 294002 inhibit autophagy in isolated rat hepatocytes. Eur J Biochem 1997; 243:240-246.
94. Maercker C, Kortwig H, Nikiforov A et al. A nuclear protein involved in apoptotic-like degradation in Stylonychia: Implications for similar mechanisms in differentiating and starved cells. Mol Biol Cell 1999; 10:3003-3014.
95. Sapra GR, Kloetzel JA. Programmed autophagocytosis accompanying conjugation in the ciliate Stylonychia mytilus. Dev Biol 1975; 42:84-94.
96. Chose O, Noel C, Gerbod D et al. A form of cell death with some features resembling apoptosis in the amitochondrial unicellular microorganism Trichomonas vaginalis. Exp Cell Res 2002; 276:32-39.
97. Chose O, Sarde C-O, Noel C et al. Cell death in protists without mitochondria. Ann NY Acad Sci 2003a; 1010:121-125.
98. Chose O, Sarde C-O, Gerbod D et al. Programmed cell death in parasitic protozoans that lack mitochondria. Trends Parasitol 2003b; 19:559-564.
99. Segovia M, Haramaty L, Berges JA et al. Cell death in the unicellular chlorophyte Dunaliella tertiolecta. A hypothesis on the evolution of apoptosis in higher plants and metazoans. Plant Physiol 2003; 132:99-105.
100. Ameisen JC. On the origin, evolution and nature of programmed cell death: a timeline of four billion years. Cell Death Differ 2002; 9:367-393.
101. Abraham MC, Shaham S. Death without caspases, caspases without death. Trends Cell Biol 2004; 14:184-193.

INDEX